OUR GENES, OUR CHOICES

HOW GENOTYPE AND GENE INTERACTIONS AFFECT BEHAVIOR

OUR GENES, OUR CHOICES

HOW GENOTYPE AND GENE INTERACTIONS AFFECT BEHAVIOR

DAVID GOLDMAN

Laboratory of Neurogenetics, National Institute on Alcohol Abuse and Alcoholism, National Institutes of Health, Rockville, Maryland, USA

This book was written by David Goldman in his personal capacity. The opinions expressed in this book are the author's own and do not reflect the view of the National Institutes of Health, the Department of Health and Human Services, or the United States government.

AMSTERDAM • BOSTON • HEIDELBERG • LONDON
NEW YORK • OXFORD • PARIS • SAN DIEGO
SAN FRANCISCO • SINGAPORE • SYDNEY • TOKYO

Academic Press is an imprint of Elsevier

Academic Press is an imprint of Elsevier
32 Jamestown Road, London NW1 7BY, UK
225 Wyman Street, Waltham, MA 02451, USA
525 B Street, Suite 1800, San Diego, CA 92101-4495, USA

First edition 2012

Notice
No responsibility is assumed by the publisher for any injury and/or damage to persons
or property as a matter of products liability, negligence or otherwise, or from any use or
operation of any methods, products, instructions or ideas contained in the material herein.
Because of rapid advances in the medical sciences, in particular, independent verification of
diagnoses and drug dosages should be made

British Library Cataloguing-in-Publication Data
A catalogue record for this book is available from the British Library

Library of Congress Cataloging-in-Publication Data
A catalog record for this book is available from the Library of Congress

ISBN: 978-0-12-396952-1

For information on all Academic Press publications
visit our website at www.elsevierdirect.com

Typeset by MPS Limited, Chennai, India
www.macmillansolutions.com

Printed and bound in the United States of America

Transferred to Digital Printing, 2013

Working together to grow
libraries in developing countries

www.elsevier.com | www.bookaid.org | www.sabre.org

ELSEVIER   BOOK AID International   Sabre Foundation

Contents

A Note on Gene and Protein Symbols

Some readers will be puzzled as to why symbols for genes and proteins (e.g. MAOA, *MAOA*, *Maoa*, *maoa*) are sometimes capitalized, sometimes italicized and sometimes not. Justifiably puzzled! It turns out that there is considerable information conveyed – for example it can be discerned whether the symbol represents a human gene versus a mouse gene or whether the symbol represents the gene that encodes the protein versus the protein itself. Although this is the type of nuance one does not have to necessarily worry about unless one wants to publish or publicly record a genetic finding, or unless one just wants to annoy people, I and the Editorial staff at Elsevier tried diligently to use the correct symbol. This is for the purpose of encouraging precision of thinking or at least not discouraging it, and to draw attention to complex but ultimately accessible levels at which genetics is done – gene, RNA and protein, for example. Also, and whether we are talking about mice or men, it is not necessarily a tragedy if our best efforts fall short of perfection, as they usually do. I suspect that somewhere in this book the wrong symbol was used and if not there are gene symbol errors in my own published work.

GENE AND PROTEIN SYMBOL NOMENCLATURE, SIMPLIFIED

For humans, non-human primates, domestic species such as dog and horse and default for everything that is not a mouse, rat, fish, worm or fly, gene names and names for corresponding messenger RNA (mRNA) and DNA made complementary to mRNA (cDNA) should be capitalized and in italics (e.g. the *MAOA* gene).

For almost all other species, mouse, rat, worm and fly, gene names should be lower case except for the first letter, and the gene symbol should be in italics (e.g. the *Maoa* gene knockout mouse). Naming of mRNA and cDNA molecules again follows the same convention as for the gene itself.

Protein symbols from whatever species except fish are in upper case and with no italics (e.g. the MAOA protein).

Full gene names are not in italics and Greek symbols are not used (e.g. monoamine oxidase A).

However, what if it is a fish?

For fish, the full gene name is italicized and all in lower case (e.g. *monoamine oxidase*).

Fish gene symbols are italicized and all in lower case (e.g. *maoa*).

Fish protein symbols are the same as the gene symbol, but only the first letter is upper case and they are not italicized (e.g. Maoa).

Why are fish different? Good question. Probably it's best not to ask how the gene or protein would be named if a fish gene is genetically engineered into a mouse, or temporarily expressed in a mouse's brain, or *vice versa*.

For those who would like to locate a gene, identify a gene by its symbol or begin their own explorations in genetics and genomics, an essential resource is:

NCBI – Entrez Gene http://www.ncbi.nlm.nih.gov/gene. This is a unified query environment for genes, DNA sequence and RNA and protein manifestations of genes. It includes names, symbols, publications, gene ontology (function), chromosome positions, genetic variations and population variation and much, much more on genes, the RNA sequences that are transcribed from them, the proteins they encode, the tissue in which genes are expressed, epigenetic features of the genome and even small molecules.

Foreword

The amazing recent advances in molecular genetics have led to several books explaining what genes "do" and how that knowledge informs us on the role of genes in influencing people's behavior. This has been paralleled by discussions of the ethical issues and of the problems involved in studying the effects of genes on the liability to poorly conceptualized mental disorders. Readers of this book will learn a lot on all of these important issues, but the book comprises a set of essays and is not an academic textbook. The main focus rather concerns the neurogenetic origins of behavior and the attendant possibilities and limits of behavioral prediction. David Goldman tackles head-on the reality of free will, arguing that this stems from neurobiology and not chance. Nevertheless, our brains capitalize on randomness as the raw material for individuality. The study of impulsivity (in humans and other animals) is used as a telling example of both the powerful influence of genes and the importance of social context, as well as the role of gene–environment interplay and the operation of epigenetic influences on gene expression.

The use of multiple practical examples and analogies, presented in a witty and engaging fashion, makes all of these complicated issues easy to understand, without in any way pretending that everything is known, which manifestly it is not. It is argued that society works best on the basis of behavioral diversity, as does evolution. The overall message is manifestly optimistic – not because of a naïve expectation that everything will turn out right, but rather because an increased understanding of neurogenetics should provide us with the means to ensure that the advances can be used for general benefit instead of discrimination and disadvantage. David Goldman skillfully avoids evangelism and ideology, whilst accepting that values will, and should, influence what we do. All the best scientific communicators make an impact through an enthusiasm for what they do, as well as through the clarity of their exposition. David Goldman is no exception to that generalization. He is a provocative writer in the best sense and that helps in making this book a really good read, as well as a stimulus to all of us to think more clearly on the multiple, exciting and important issues involved in neurogenetics.

Sir Michael Rutter
Professor of Developmental Psychopathology
King's College
London

Preface

The book you are reading is a story, and an analysis, of the neurogenetic origins of behavior and free will in the context of a modern appreciation of neurogenetic determinism. None of us chooses our parents, our ancestors, or the past world that shaped our genomes and that in the present constrains our behaviors. From early in our lives we may hear "whatever will be will be". We are also told that the life we make and what we become depend on ourselves. From an early age we learn to treat people as if they are free. We reward or punish them *as if* they can control their actions, and we accord to them the gift of respect as autonomous people, rather than treating them as *things*. In part we do this because we can empathize with their thoughts and emotions, and understand the conflicting motivations that led to the behavior that we may or may not like. The failure to make such an emotional connection turns other people into objects and, as they say, makes the world a little colder. However, what if a judge is confronted with a genetic finding that a murderer carries the "2B or not 2B" variant, a human polymorphism found by my own laboratory that predisposes some individuals to impulsively murder? Should that information be used to mitigate the sentence or, paradoxically, should the carrier of this gene be viewed as even more dangerous and likely to repeat the act?

In writing this book I was inspired in part by response to lectures at national judges' courses on the validity and meaning of such evidence. DNA and brain scans are increasingly used in courtrooms. What should be allowed? How should it be weighed? Is there a neurogenetic basis of moral responsibility, or do we merely treat people "as if" they are free? Personal genomics includes the new ability to sequence genomes, raising issues of how the information will be used, as illustrated in popular movies such as *Gattaca*. Thousands of people purchase genetic scans to better understand themselves. However, the reports they receive are sparse and subject to misinterpretation. People are intensely interested in what their genes say about their personality and behavior. Therefore, I wanted to make this book as "user friendly" as possible. I tried to minimize the jargon but it is hoped that the glossary at the end will assist people who get stuck on some piece of terminology. Also, instead of stringing together a series of stories, I strove to select stories that could illustrate a step-by-step account of how our DNA builds a brain, and how in the end it builds a brain capable of freely choosing whether to read all or any portion of a book like this.

Spectacular advances in genetics and genomic science have led to the reading of the genetic code of our species, and the identification of genetic variants that alter human behavior in ways that are profound and surprising. As illustrated by the story of the "2B or not 2B" gene, the mystery of whether people are free has only been deepened by the new perspectives on neurogenetic determinism that are described in this book. As a neurogeneticist, it is a pleasure for me to chronicle some of the most interesting and profound ways in which genes influence human behavior, from the rich and strange new landscape unveiled by the genome revolution. Reflecting our evolutionary community of relationship with other life on Earth, many behaviors, for example our sexuality, are paralleled in other species and some are even influenced by genetic variations that act in similar ways and with similar effects. As I like to say, it is my hope that something that we learn about the origins of behavior of people will help us comprehend the elusive behavior of cats. In people, these discoveries are giving us new insights into the origins of psychiatric diseases, and as I tell the story, the ways that genetic variation interacts with environmental stress and even endocrine differences to make some people "warriors" and others "worriers". We are neurogenetically individual, but does that mean that from conception our DNA is destiny or, as I will argue, is our DNA heritage paradoxically an essential ingredient in the freedom of the individual?

Life comes full circle, and in my case the nature of the boy was the blueprint for the man. I became a neuroscientist because as a child I perceived that our brains contained the essence of self and humanity, and there could be no better way for us to know ourselves than to study our brains. However, when I was about five years old I perceived a problem of which I have remained acutely conscious ever since. All life is based on causal relationships in a physical universe. The more I learned about myself and others the more apparent became the connections within webs of causality. Could freedom exist in a world where things do not happen by magic? Why was it that as one of five children, I was like but also unlike my older sister, my younger brothers, and my parents? My dear brother Paul developed schizophrenia, and later enriched the lives of all in the family with his weekly letters on diverse topics. Why? What were the factors in his genetic predisposition, and what were the wages of his experience? Why do psychiatric diseases, and as I will describe, terrible acts such as murder, strike close to home and blight the lives of so many of our fellow humans, who we would also spare such travail?

The genomic era holds unprecedented promise for the discovery of clues to the origins of common, genetically influenced psychiatric diseases such as my brother Paul's, and hopefully will contribute to their prevention and treatment, and perhaps even interrupt the sequence of causality within those who would victimize others, the victimizers themselves being a type of victim. We are only at the beginning of the beginning of an exhilarating process of discovery in understanding the role of genes in the developmental program of the brain. The way the mystery is unfolding, and my own ability to make some contribution, is everything I could have hoped for when I first arrived at the National Institutes of Health in 1979 from Texas as a raw and overly optimistic young physician with a burning desire to understand the human brain and make a name for himself. The progress is also frustratingly less than what is needed for families such as my own and for the millions of patients and families fighting their own everyday struggles against psychiatric diseases. Severe psychiatric diseases, whether schizophrenia, depression, alcoholism, other addictions, obsessive compulsive disorder, attention deficit hyperactivity disorder, autism, bipolar disorder, phobias, panic disorder, eating disorder, borderline personality disorder and antisocial personality disorder (and the list goes on) are common and affect someone in nearly every family. Yet all are stigmatized to one degree or another and many are frequently denied. For each of these diseases, therapy is only partially effective. Several have been the particular focus of an anti-psychiatry movement that would deny that these problems should be classified as diseases or that people with them can benefit from medical treatment. And, indeed, it is not uncommon that treatment is inappropriately applied.

However, even with our crude knowledge of psychiatric diseases, and a correspondingly pitiful number of diagnostic categories, we can be sure that genes play a strong role in causation. This is because of the inheritance of these diseases. Heritability = Genes. This book describes some of the first instances of gene discovery, and false discovery. I forecast what may soon come to be, as the next generation of scientists applies genetic tools immeasurably more powerful, for example using the new ability, described in this book, to measure the combined imprint of genotype and experience on the genome. Also, we can already be certain of the role of environmental interaction, not only in terms of the general role of stress and endocrine factors, but via the specific gene by environment interaction stories that I tell.

This book describes how we know about the relationship of genes to behavior, which are foundations for any theory of behavior or public policy. It also describes a most exciting aspect of the story, which is that as we penetrate deeper into the brain's secret functions, we find that actions of genes tend to be progressively stronger. At the same time, many genetic variants altering human behavior are not "disease genes", but are literally genes "for" behavior, having been selected not to alter some obscure brain function but to influence the behavior itself, leading to differences in personality, cognition and reaction range to the environment, including stress and addictive agents. This simple fact, and the fact that we all carry thousands of genetic variants, some beneficial and some highly deleterious, represents a profound refutation of the eugenic impulse to identify some humans as "fit" and others as "unfit". Fitness is all about context.

Behavioral genetics has a long and productive history, although it was marred early in its course by eugenics, and polemics on gene versus environment and racial differences, as this book discusses. Behavior geneticists quantified the inheritance of behavior and psychiatric diseases. However, with the completion of the draft sequence of the genome, neuropsychiatric genetics has leapt forward to the level of the specific genes and gene variants that alter behavior and how they work. This book tells the story of how this research is being conducted not on a gene-by-gene basis, but on a global, which is to say "genomic" basis. We have an exciting first glimpse at what we are, and why we are, and the results are startling and controversial in many ways. In modern scientific parlance, the old paradigm of understanding human behavior has been broken or "disrupted" and now our field is struggling to replace it with something new, and based not on a quantitative appreciation of causality ("people who experience stress tend to become depressed") but on an more exacting understanding of who that person was before and after the stress.

For me, it has been a privilege to be part of this first generation of these "behavioral genomicists". They are a variable and fascinating lot, and comprise the cast of characters in this book. I hope that I have not done any of them a disservice with how they are represented, or by omitting them. I have mainly written about fellows, friends, colleagues and scientific heroes who I personally know. Some were in my own laboratory, some in laboratories with which I collaborated or competed against, and some are just old friends and competitors, with whom I have talked, argued and written. It would not be "in me" or "like me" to write from a biographical perspective, for example via systematic study of letters or interviews. Yet it is important to point out that in mentioning people in connection to one piece of science, shared events or something memorable they said, it was not my intention to minimize or encapsulate their lives or contributions. I only know what I know, and it is from that perspective, as a physician scientist working as a Laboratory Chief at the National Institutes of Health, that I wrote this book. It also might go without saying that this book is not an autobiography. Much more remains unsaid than could be said, even if there was "more space in the margins". Meanwhile, the stories I selected are directed towards one fascinating question: Are people free to make choices, or do genes determine behavior? Choosing the stories was a main enjoyment of writing this book. Neurogeneticists are spoiled for choice, and yet I have seen that most of these stories are scarcely known to the general public.

For example, in some fish, a single genetic switch causes a change from female to male behavior. Humans are more complicated, but all human behaviors emerge from the expression of DNA. Pathways from DNA to behavior are illustrated via new advances, for example:

- In people, is there a "gay gene"?
- The first genes that predict cognitive and emotional differences.
- The ability of genes to alter behavior.

Interaction between genes and the environment is explained via human genetic variants that alter depression, anxiety and impulsivity.

Are people free to make choices, or do genes determine behavior? My studies of human behavior convince me that the answer to both questions is paradoxically "yes", because of neurogenetic individuality, a new theory with profound implications. In *Our Genes, Our Choices*, the complexity of human behavior and a person's ability to choose are explained as deriving from the ways in which a relatively small number of genes direct a neurodevelopmental sequence. This lifelong process is guided by individual genotype, molecular and physiological principles, by randomness and by environmental exposures that we choose as well as ones that we do not choose. This theory affirms and provides a mechanism for the origins of free will and the ability to make moral choices, but I suspect that the debate will continue as long as humans exist.

Dr David Goldman
Director of the Laboratory of Genetics
National Institute on Alcohol Abuse and Alcoholism
Rockville, MD

About the Author

David Goldman is one of the world's most influential neurogeneticists, with more than 400 publications cited over 22,000 times, many published in the world's leading journals. He is at the forefront of revolutions of imaging genetics, gene by environment interaction, pharmacogenetics and deep sequencing to identify genes that alter human behavior. He founded the Laboratory of Neurogenetics in 1991 at the National Institutes of Health. He has received many awards, most recently the NIH Director's Award, presented by Francis Collins for elucidating gene by environment interactions in alcoholism and other behaviors. He is past Chair of a human research review committee and has often spoken on neurogenetics, including on genethics. Born and raised in Galveston, Texas, he attended Yale University, where he graduated in three years. He received his MD degree Magna Cum Laude from the University of Texas Medical Branch. He has three children and is married to Nadia Hejazi MD, a pediatric neurologist. Since 1985 he has lived in a house of his design in the woods in Potomac, Maryland, and commutes by bicycle to the NIH throughout the year.

The Neurogenetic Origins of Behavior

"תמשל"

Transliterated as "Timshel": Thou mayest choose

Genesis, Chapter 4, verse 16

Is the human will free? Do genes determine behavior? Paradoxically, the answer to both questions is "yes". A new theory of behavior based on neurogenetic individuality has profound implications for conceptions of self, social expectations, ethics and justice.

This book begins with a challenge to free will from my research. The discovery: "2B or not 2B?" involves a gene "knockout" of the human *HTR2B* gene, which encodes a receptor for the neurotransmitter serotonin. Serotonin is involved in many aspects of behavior including emotion and impulse control. The severely functional receptor variant causes some people to be impulsive, even to the extent of committing senseless murders. Remarkably, it is found in at least 100,000 people in the Finnish population, but as a "founder mutation" has so far only been observed in individuals who are of Finnish ancestry. Yet, while the inherited variant was a "necessary" factor in the impulsive murders that I and my partners in research studied, the gene alone was insufficient to explain the heinous behavior. In understanding why the carriers of the receptor gene variant became murderers, "2B or not 2B?" was *not* the only question. The context of the gene, for example male sex and drunkenness, also mattered.

All human qualities, including those that are sublime, creative and adaptive, and those that are seemingly mundane, destructive and maladaptive, are ultimately emergent from the expression of a "message in a molecule". That molecule is DNA. DNA is an information molecule – a polymer in which information is encoded. For example, our DNA contains some 25,000 genic protein-coding regions. However, DNA is ultimately only a chemical that is now fairly easy to synthesize in the laboratory. The total DNA of a bacterium was recently made in a laboratory,

Our Genes, Our Choices
DOI: 10.1016/B978-0-12-396952-1.00001-4

and it is probably only a matter of time and motivation before someone synthesizes the whole genome of a more complex creature – even a human. Also, because of the tools now available to study DNA and the ability to study its effects in powerful contexts, including in animal models in which genes are knocked out, the science of genetics has now achieved what seemed impossible only a few years ago. We have demonstrated the causal connection, and not just the correlation, between single genetic variants and complex human behaviors. As will be shown, this reductionistic explanation of human behavior is only in its infancy, and complicated by many difficulties and some false leads.

With some 25,000 protein building blocks and probably an equal number of regulatory RNA molecules, and even allowing for variations in structure, how is it possible that the DNA message can encode a human brain, with its 10^{15} (one billion × one million) connections? How, based on the DNA code, can a brain build itself? Is it possible that the complexity of human behavior, and even free will, can be derived from the chemistry of DNA? As will be proposed, the answer lies in the way that this relatively small complement of genes directs a developmental sequence that continues throughout life and that is guided by principles, stochastic in countless details and always completely individual.

Pathways from neurobiology to complex behavior, and the ability of "things we cannot control" to shape behavior, are illustrated by sex. Why do men and women behave as if from different planets and, beyond the effects of culture, how are we to understand the origins of variations in sexual behavior of sex-specific behaviors ranging from attachment, to aggression, to homosexuality? Is there a "gay gene"? Why is there a genetics of sexual behavior and how can we understand the diversity, or even perversity, of human sexual behavior? A clue is that in other animals there are sex genes and several have been identified, although in the human the identity of these genes is yet unknown. What are the implications for choice and conceptions of personal freedom that people are born male, female, or gay, or that a switch in the function of a single gene can cause a fish to stop acting like a female, and start acting like a male?

Despite the complexity, and using methods that can find the needles hidden in the genetic haystack, I and other neurogeneticists have identified the first genes that predict cognitive and emotional differences, and sometimes the same gene can have countervailing effects on both. The ability of genes to predict behavior is explained by example, as are the limitations and nuances, which include gene by environment interaction. The genetic variants identified so far are not strongly predictive of behavior and their value should not be oversold, as has already been done. Yet they are of explanatory value, and foreshadow the discovery of additional variants that must account for the heritability of human behavioral characteristics.

The new science of gene by environment interaction is explored via genes that can lead to depression, anxiety, impulsivity and even suicide. For example, genes predispose some people to be resilient "warriors" and others to be less resilient and pain resistant but sometimes cognitively advantaged "worriers".

The findings in this relatively young science of behavior genetics are not without controversy, as would be expected. However, there appear to be several solid examples of genetic variations that affect behavior, and also affect what the brain is doing during behavior, as is now observable with brain imaging, which is a window on the activity of the brain. By imaging the regional structure, activity and chemistry of the brain, for the first time the basis of the effects of these behaviorally important genes has been understood. Also, their predictive effects on brain function itself is much stronger than on overtly manifested behaviors. Several genes with weak effects on anxiety and emotion have strong effects on brain responses to emotional challenges. Two genes that influence cognition have strong effects on brain activity while people are asked to perform cognitive tasks that challenge specific parts, and neuronal networks, of the brain.

This book is concerned with the neurogenetic origins of behavior and with the possibilities and limits of genetic behavioral prediction; however, it is unavoidable that these discoveries would be connected to conceptions of self and freedom. Is it sufficient to treat people "as if" they have free will? I argue that it is not. The "as if" stance is inherently inconstant, setting the stage for the easy erosion of individual autonomy whenever situational ethics dictate that it is more expedient to treat people as slaves to causality. *Compatabilism* is the philosophical position that determinism and free will can be reconciled. However, I would also take issue with Daniel Dennett's compatabilistic formulation, which holds a dependency on culture, "morality memes", and child upbringing, and even the idea that to have free will one must believe in free will or be disabled as a chooser. All are metastable foundations. As will be discussed, individual and group autonomy are the vital bases of moral ethics, for example as applied in the conduct of human research, where these are foundational principles. However, are these principles divined from a philosophical or practical calculus or are they inherent to human nature? Are they suppositions, in which case any system of moral ethics may define humans otherwise, or are they parameters based upon observation?

I will argue that individual free will, and by extension the autonomy of groups of people, are parameters whose existence can be derived from the inheritance of cognitive structures and variation in these structures due to developmental neuroadaptation. Returning to the question of whether choice can emerge from a brain shaped by genetic and environmental determinants, a conception of neurogenetically determined free will is unveiled that at first seems paradoxical if we are only wading in the shallows of neurogenetics, where the focus naturally gravitates to how genes and environments influence the behavior of groups of people. A deeper analysis reveals that each human, including an identical twin, is neurogenetically individual, and that brain development unfolds stochastically throughout life in a way that makes each of us a unique and ultimately self-determined entity. As a consequence, each of us is also unpredictable, but it is not this unpredictability that represents freedom. Our freedom is bound to our individuality, which is partly the product of neurogenetic determinism, which is itself bound to the ways the human genome, and brain, were shaped by evolution, and that would include the random events that altered humankind's evolutionary path. Freedom does not consist of randomness: philosophers such as Robert Kane and Daniel Dennett are correct to emphasize that free will does not originate in quantum randomness. However, our brains capitalize on randomness as raw material for the development of our individuality. Dennett has warned that we should not look too closely at our mental activities or we may discover that we have no selves. Coming at the problems of self and free will from a neurogenetic perspective, I am contending that it is these biological parameters that define our individual, free selves. As we solidify our understanding of genetic and environmental predictors, behavioral prediction inevitably improves – we can better anticipate the responses of any particular person, be they neuroscientist or philosopher or someone intrigued by their musings – but whether or not someone may guess our choices, we may choose.

The Jinn in the Genome

FIFTEEN MINUTES OF FAME

In 2002 I helped to conceive "Our Genes/Our Choices", a Public Broadcasting series that explored ethical and legal choices created by the genome revolution. Topics of these Fred Friendly seminars included genetic reproductive decisions and genetic privacy. Our session, "Genes, Choices and the Law", was adeptly moderated by Charles Ogletree, a well-known Harvard Law Professor. The scenario involved an alcoholic accused of a crime, and a genetic test result that might be mitigating. I was the geneticist with a predictive test. My molecular geneticist partner on hand was Dean Hamer, already celebrated for having discovered the "gay gene" (more on that later). Little did I know what I was getting into, and as they say, my 15 minutes went by so fast. Seated on my left was my "boss", Francis Collins, who later indeed became Director of the National Institutes of Health. On my right was Justice Stephen Breyer.

Our Genes, Our Choices
DOI: 10.1016/B978-0-12-396952-1.00002-6

Other superstars included lawyer Nadine Strossen and journalist Gwen Ifil. Representing the defendant was Johnny Cochran.

The case was a puzzle in genetics and the law. Can genes predict behavior? Should predictive tests be used and if so how? What about genetic tests that could stigmatize groups of people? If a judge rules that evidence can be introduced that a gene influenced criminal behavior, would identification of this genetic link in a chain of causality influence our ability to convict someone, or modify the penalty? Several people including Dean Hamer suggested that I write a book about genes and behavior. However, at that time (and while Dean was writing another) I was grappling with a foundational issue with which the puzzle of genetic prediction is bound. This was the problem of determinism and free will, and the implications of inborn genetic determinants for behavioral choice. If I had not made up my own mind, what business did I have trying to change someone else's? Eight years later, I thought I might be ready to attempt a new synthesis on human choice, based on a concept of neurogenetically determined behavioral individuality.

SOME FAMOUS GENETICISTS, AND WHY THEY ARE FAMOUS

As has been well chronicled, the draft sequence of the human genome was published in 2001 by two rival groups, a corporation led by Craig Venter and a government consortium led by Francis Collins, who co-discovered the cystic fibrosis gene and who at that time directed the Human Genome Institute. Because the race for the sequence ended in a virtual dead heat, it was appropriate that both were honored by President Clinton in a Rose Garden ceremony. However, as Francis Collins observed, completion of the draft sequence was only the end of the beginning, bringing us to the starting line of a much longer race to understand how the genome works, and to prevent and cure diseases. The advances keep coming. Last year, Craig synthesized the complete genome of a bacterium. This in-laboratory duplication of nature underlines the fact that all life on Earth, and its complex variations, is ultimately based on the expression of complex chemicals: DNA and RNA, and those chemicals are increasingly open to measurement and manipulation. Beyond its biomedical applications, knowledge of the human genome may enable us to answer some of the most fundamental questions as to what humanity is and to what peaks it might lift itself, by its own bootstraps. If genetics can never provide all the answers, it can at least do what science does best, which is to facilitate the asking of better questions and new questions. Also, as will be discussed in some detail, genetics properly applied has a nearly unique ability to establish causal connections, and not just correlations, between the molecular level of the DNA code and people's most complex attributes, including their behavior.

James Watson, who with Francis Crick won the Nobel Prize in Physiology and Medicine for deciphering the structure of DNA, conceived the Human Genome Project. Watson recognized the critical importance of *genethics*, and as Director of the Human Genome Institute he therefore set aside a fixed percentage of the funding for the ethical and policy implications of genomics research. This ELSI (Ethical, Legal and Social Implications) program has had a lasting and pervasive impact on the thinking of human geneticists, in part because many, including me, participated in symposia it sponsored. From the beginnings of human genomics research, scientists grappled with the societal implications with some important practical results including the passage of GINA, a federal law that will make it more difficult to discriminate on the basis of genetic information, as will be discussed in more depth in Chapter 9.

However, it is also fair to say that the impact of genetic knowledge and of the powerful new tools that implement it remains a work in progress for the scientific community and has scarcely been appreciated by the public at large. As the leading edges of genomic knowledge and technology rapidly advance and science becomes more specialized, it becomes more difficult for generalists to make accurate assessments, and to some extent experts are being asked to predict the future. For example, will stem cells made from adult tissues suffice for transplantation medicine, and in what time frame and at what costs? The important advances in genetics are coming from scientists with very widely varying goals, perspectives and backgrounds. Some are human geneticists and some are not. Some are physicians and some are unfamiliar with medicine. Some have had direct involvement in the medicolegal side of genetics and others not. The public and the scientists advancing frontiers of knowledge will repeatedly be presented with new ethical puzzles created by new capabilities. Making wise use of the knowledge will require a continuing reevaluation and readjustment of mindset, and the input of many voices.

THE JINN OF KNOWLEDGE AND THE JINN OF TECHNOLOGY

Everyone is aware that there has been a genome revolution, but probably all but a few visionaries and writers of science fiction underestimate the implications. By studying our own genomes – the genetic blueprints of our lives – humanity picked up a lamp and two powerful forces were released. They wait expectantly, and we should be very careful what we ask of them. Unlike "real" jinns, they are at large in the world and answer to no one master. Genomics knowledge and tools are available in any country. We cannot assume that the decisions we make will determine how these tools are used elsewhere, but we can influence those decisions by word and deed.

The jinn of knowledge is the linear sequence of the human genome discovered by the Human Genome Project, and soon we will have much of the genome's three- and four-dimensional depth – its variation between individuals and populations, and its complex regulation including the unfolding of the DNA-encoded developmental programs that enable us to develop from a single cell into a complex organism. The jinn of technology is the ability to rapidly, accurately and cheaply measure any genome and its functional outputs. Applying the new technology and the template of genomic knowledge, we shift our level of evaluation of personhood to the level of molecular predictors. Obviously, this molecular view of the person is particularly potent prenatally, postnatally and in infancy, when the individual has had little opportunity to establish their own attributes. We cannot know exactly how the infant will develop but at birth or prenatally we can obtain a blueprint that is increasingly predictive. It will be tempting to use it and for some purposes we must use it. In adolescents and adults the genome sequence, as well as marks of the environment on the genome (the rapidly growing science of *epigenetics*, which will be introduced), is also informative – it helps us to understand how the individual came to be what they are, and can help to predict their future responses. It is a purpose of this book to explore the uses of this predictive power in behavioral genetics, for better or worse.

REVOLUTIONS IN CULTURE AND EVOLUTION OF GENES

We live in a progress-oriented society, and indeed civilizations that fail to embrace change and that languish technologically are likely to be consigned to the sidelines of history, if not obliterated by another culture with better guns, productivity or, in the Jared Diamond sense, sustainable systems for living.

The problem of genome technologies and other powerful and transformative tools is a dilemma of change. Other societies may embrace what one forgoes. It is therefore unsurprising that in all the more successful modern societies in which books such as this one are written on electronic media there is collective agreement on an ideology of progress (consider the pompous corporate slogan, "Progress is our most important product"). The need to embrace technology explains much of the behavior of contemporary nations, and their relative success. Well before an atomic bomb obliterated Hiroshima and a beeping Sputnik awoke Americans to the fact that they lagged in the space race, it was widely recognized that the most important natural resource of a nation is not oil or uranium, but its populace. How large is it, how well educated and how well enabled to compete? The space race launched by Sputnik ultimately helped to end the Soviet Union, but as a nation rich in intellect, Russia is again rising and one possibly beneficial side effect is that some of its bright intellects migrated around the world. The economic engine of China has been unleashed by leaders raised under communism but wise enough to comprehend that it was not working. As exemplified by their pains with the internet, the need of the Chinese to join in the global interchange of information, so well exemplified by the diaspora of Chinese people, will inevitably erode authoritarian limits on speech.

Overall, scientific advances make us better off, but whereas change is inevitable the devil lurks within the details of application and laws of unintended effect. Things usually do not unfold exactly according to plan. However, even when they fear the consequences, scientists are drawn to experimentation, invention and mastery. In Kurt Vonnegut's *Cat's Cradle*, ice-nine (a water crystal of high melting point) froze the world's oceans. In the real world, the first atomic bomb was detonated in a desert called *Jornado del Muerto* (Journey of Death). Observing the unmatched force, Robert Oppenheimer said, "Now we have become Death, Destroyer of Worlds." But does it really improve matters to destroy the world while making sophisticated comments that will be memorable only if anyone survives to hear them? Were those "just pretty words"? As the composer John Adams asked in *Doctor Atomic*, what were the physicists really thinking and did they see no other way out? For better or worse, I believe that Edward Teller, the father of the superbomb, the hydrogen weapon, was certain of the correctness of his actions, but Oppenheimer was tortured by the potential for future catastrophe created by an unparalleled power in human hands.

Fortunately for life on Earth, the first fission blast did not trigger a thermonuclear chain reaction in the planet's atmosphere, as some physicists had feared. It led to the death and disfigurement of thousands of civilians in two cities and to a new and terribly dangerous era that continues to threaten life on Earth. Oppenheimer was a tragic figure precisely because he foresaw some of the things to come, and even the possibility of Armageddon. Working in the realm of imagination, in *Slaughterhouse 5*, Kurt Vonnegut dreamed up the Tralfamadoreans, an alien race gifted with transcendent foresight into the mists of the future. However, the Tralfamadoreans also had a fascination with the fundamental structure of matter and the design of an "ultimate weapon". At a point in the future – now imminent – their project approaches completion but beyond that point the Tralfamadoreans cannot see, possibly because they have destroyed the universe and there is nothing to see. Faced with the ultimate choice they continue to do what it is their nature to do: build the weapon. Are humans like Tralfamadoreans in their inability to resist an impulse to open the next, and deadly dangerous, "black box"?

It is fair to say that we scientists spend our lives opening as many black boxes as we can get our hands on, and are usually either not thinking about the consequences or rationalizing. There is a little Tralfamadorean in all of us and many of the activities of science and technology expose humanity and the world to dangers that are quite unexpected. We are not prescient. When fire was captured by man, no one would have thought that the burning of hydrocarbons would one day alter the climate of the whole planet. When lead was added to ceramics, gasoline and paints, no one expected that humanity would one day pay a price for its toxicity, and it took generations of exposure and observation to unravel the causal connection.

Yet, scientific knowledge can be our greatest shield against consequences. Tralfamadoreans, Oppenheimers, Tellers and their physics experiments aside, the danger is rarely the experiment itself, it is the technical application of the knowledge. It is science that enables us – to the limited extent and ability humans have – to look into the future and understand the consequences of our actions. It is because of science that we understand the effects of heavy metals and toxins in the environment and have used that knowledge to save lives and improve lives. The developing brains of children are shielded from lead that would otherwise diminish their capacities. Similarly, those of us who are fortunate enough to live in societies where the knowledge is applied are routinely shielded from a multitude of hidden dangers ranging from the inorganic such as asbestos to the organic such as vitamin deficiencies, viruses and bacteria.

Science gave us the crystal ball that predicted the consequences of the release of fluorinated hydrocarbons into the atmosphere, and that knowledge enabled a successful international effort that has preserved the ozone layer, which is now recovering from its depletion. The crystal ball is often a bit cloudy, but it is because of science that we know that the massive release of carbon dioxide into the atmosphere is causing global warming. Many individuals who know – and who do not deny – are constantly working in large ways and small ways to at least mitigate the potentially catastrophic effects. As for me, I will at least commute to work on a bicycle until some motorist pries my cold, dead fingers from the handlebar. Without science, global warming and these other catastrophes are threats that are as invisible as they are potent, but with science we can see that they are in our possible futures and can make rational choices to avoid them.

However, there is a mismatch between progress in genetic and other scientific knowledge and technological progress. This lag between technological progress and our knowledge, or the ability to integrate and use that knowledge – which is the essence of wisdom – is what endangers us. We become aware that knowledge is unequally distributed and it is frustrating when others seem to be unaware of important facts. Then we become aware that the problem is not just these "other people", with whom we disagree politically. We have begun to lag behind the knowledge curve in many important areas. A parent may appreciate this when watching the facility with which their children grasp the latest technical modality.

The new generation is always better able to adapt, but ultimately as a species, how well do our genomes match up to the new challenges? Can our brains and scientifically based understanding enable us to properly use, balance and adapt to the inventions and consequences of our science?

Perhaps not. Our handheld electronics and fiberoptic cables serve us better than stone flints and steel. However, the demands that the world we call modern places on humans are arguably not only different but more difficult than the demands that were presented in previous millennia. Humankind's first steps up the cultural ladder required genetic changes enabling language and complex social behavior, a process that took millions of years. In the past 60,000 years, which is almost the last instant of time in human evolution, technological progress has advanced exponentially and has become decoupled from genotype. The technological revolutions – agricultural, industrial, atomic, computer, communications, internet, genetic, nanotech and biotech – erupt at shorter and shorter intervals, as shown in Figure 2.1.

As the pace quickens such that many people have actually experienced seven technological revolutions within their own lifetimes and, to a considerable extent, as the nature of the technological "race" changes, many fall behind. Email was a wonderful tool until we discovered that it turns "work" (whatever that is) into a 24-hour-a-day proposition. Cell phones are handy but are there not some times when each of our lives would be improved by an interval of silence, for perhaps the space of half an hour?

Life is a continuing education. The Red Queen, a role my technologically advanced niece Hannah performed, summed up the challenge, "Now, here, you see, it takes all the running you can do, to keep in the same place." Our genomes cannot keep up. However, our brains and our networked brainpower can. By knowing ourselves and the world better, and integrating that knowledge, humanity can make wiser choices. "Wiser" means decisions informed by science and knowledge of the world and humanity itself. "Choice" means free decisions made

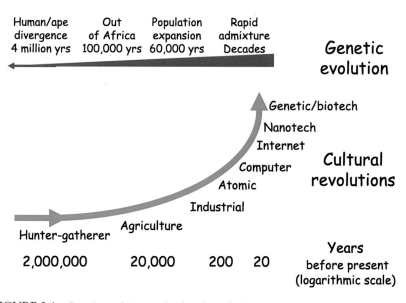

FIGURE 2.1   Genetic evolution and cultural revolutions.

by individuals and individuals acting in groups. This book, as a tour of behavioral genetics – where has it gotten us and where is it going? – is an attempt to contribute to the dialogue of the question and response of what we are and how can we choose, and indeed whether humans can choose. At a deeper level it advances a new concept of neurogenetic individuality and free will.

It is my belief that an understanding of the origins of behavior – our neurogenetic individuality – opens a passage to a better relationship between persons and society. The message undercurrent is that people can maximize their own potential and more adroitly surf new technological waves, and perhaps avoid a few of the most dangerous, using the genomes we have and not necessarily the genomes we might want. The message is that we are not motes in a stream or interchangeable parts in a social machine, but individuals with capability, and therefore responsibility, to choose the type of world we would like to live in and pass on to our children.

GENES, BRAIN AND INDIVIDUALITY

This book mainly explores the phenomenon of genetic variation and its effects on the brain. The sequencing of the human genome and dramatic technological innovations are naturally leading to the identification of genes that encode variation in behavior and cognition. These advances include the first keys to the inherited origins of psychiatric diseases; other behaviors that may or may not be defined as clinical disorders, depending on which generation of diagnostic categories is in use; and predictors of behavior. Even the politics of racial identity has been battered as people have come to understand that they have a unique ancestry and combination of culture and experience. This genetically and neurodevelopmentally informed perspective leads to a new perception of personal identity that transcends simplistic profiling.

The new conception of individual identity, so strongly exemplified by modern politicians, entertainers, athletes, artists, businesspeople and scientists who transcend crudely formed and often cruelly applied classifications, is powerfully based in new findings on genetic similarities. We have learned of the close genetic affinity each person has to all other humans anywhere, with only a small fraction of our heritage of genetic variation being assignable to differences between populations or races. However, our genetic variation, including at least 22 million relatively common polymorphisms, combines in myriad ways to make each newborn something the world has never seen before. Even genetically identical twins, despite their resemblance for physical and behavioral characteristics, are truly unique at all levels from molecules to cell, body and behavior, and this is true because of the stochastic (random) nature of the processes by which development unfolds.

THE NEUROGENETICS OF DETERMINISM AND FREEDOM

The identification of neurogenetic determinants of behavior suggests a new opportunity to interpret the role of volition or "choice": the key question being, "Do we have free will?" versus the more modestly framed and sometimes, but not always, practical, "Does it make sense to treat us as if we are free?" We will take on the question of whether genes can predict criminal behavior, and the one-word answer is yes (but of course it is more complicated). We will differentiate the ability to predict a thing from the origins of the thing. We will discuss the genetic origins of sexual behavior. We will see that genes can strongly influence resilience, emotionality and cognition, and that when views of the activity and chemistry of the brain are obtained these gene effects are more profound.

Although the genetic revolution is changing our understanding of all aspects of the human condition, the focus of this book is the intersection of genetic individuality and the origins and individuality of behavior, a concept that I label "neurogenetic individuality". Nearly every complex human behavior: personality, intelligence, criminal and antisocial behaviors, sexual behaviors, addictions, depression and anxiety disorders, eating disorders, schizophrenia and manic depressive disorders, is moderately to strongly influenced, or determined, by genes. However, most of these gene effects are context dependent, a phenomenon known as gene by environment interaction. Furthermore, and crucially, the genes – especially genes that alter human behavior – also create their own contexts, a phenomenon known as gene by environment correlation. Humans have an unmatched ability to seek and create their own environments.

In our innate ability to select and alter our environments and thereby to manipulate gene by environment correlations, leading to a unique and unpredictable unfolding of development of our brains, we behave as if we have free will. Thus, with all our limitations and although we live within an overall universal framework of causality, people should be treated as having free will. The genome may at first appear to be a linear sequence, like the tape read by a player piano, and with only four notes: A, G, C and T. However, this first impression is deceptive. Because of the processes by which our genome evolved and the things it evolved to do, and because of the mechanisms by which the genome guides but never programs neurodevelopment, dignity and autonomy are paradoxical emergent consequences of a linear sequence of only three billion letters.

"Why does the drum come hither?"

Hamlet, Act 5, Scene 2, William Shakespeare

The foundations of science are the monuments and rubble of past achievement. When the science of today is viewed from the perspective of the future it will be seen as small – "little did they know" – and ramshackle, with critical gaps and errors in its architecture. One goal of modern genomics approaches is to move from the piecemeal to the complete by studying the whole of the genome to understand its structure, variation and expression. This approach has paid great dividends, and also faces some ultimate limitations, mainly because the level of human genetic variation is nearly unlimited and the complexity of genomic expression is unlimited.

Although science often advances by serendipity and leaps of insight, the edifices of the most successful scientific narratives are constructed piece by piece. The way a discovery was made is, if we can get the real story, a truth, but the way a story of discovery is told is a matter of style and strategy. The step-by-step approach to storytelling is perfectly good for those patient enough to wait, but as we will be discussing, not all of us are inherently so patient. Perhaps you are one of those people who has the habit of quickly flipping through a book to the end, to see if the whole exercise leads anywhere interesting. To the consternation of some, I assay the quality of thought and prose via a quick read of several random pages somewhere in the middle. However, I do not read the ending, preferring to test my ability to guess or predict it. It is so much more delightful to guess the ending if it then takes Sherlock

Holmes 200 additional pages to figure out that the Indian snake handler, and not the parlor maid from East London, put the rare poisonous snake in the victim's bed. Furthermore, if in the first scene a gun is hanging over the fireplace it may be pointless to pretend there is any suspense as to whether it is going to be used, but an intriguing mystery in the who and the why.

Laboratory science is step by step, for sure. It is the execution of many carefully planned steps. In genetics, like cooking, we use all these good recipes (the well-defined experimental protocols), cookbooks (the handbooks of molecular biology), equipment (my lab even has a microwave oven) and ingredients made by others (the enzymes, the reagents and even the buffer solutions), and many forms must be obeyed or we simply waste effort. We try not to deviate from the recipe but when we do occasionally something amazing is observed. In molecular genetics, a scientist often performs dozens of arcane tasks to answer one question. However, a good scientist tests the predictive validity of the whole idea (the hypothesis) as quickly as possible. Instead of validating each method – a task of years, and the procedures change – we assemble a repertoire of methods that work, and these are applied as far as possible in parallel, not in series, because "science cannot wait". Also, if there is a doable critical experiment that could invalidate, which is to say falsify, the hypothesis, I tell my postdoctoral fellows to please, please Do It. Who wants to spend a career in science on what is essentially an intellectual dead end? If idea number one is wrong, move on to plan B. Much scientific research is sponsored by the US government. Scientists who dilly-dally or, as happens all the time because it is more difficult to avoid, do work that is redundant, shoddy, irrelevant or infeasible, waste not only their own opportunities but also taxpayers' money. Early in my career I was assigned a desk next to a capable scientist who was always planning and replanning an experiment he was incapable of performing. Reexamination of "the plan" always revealed a flaw. This was the experiment that could have falsified the idea.

In behavioral genetics we can list many obstacles to the discovery of genes "for" behavior, but my main response to the conclusion that it "cannot work" is to test the idea empirically. In my opinion, several genes for complex behaviors have been identified, which would move the discussion from whether we could discover genes "for" behavior to how many we could discover, and to the implications of what we find. Much of the focus of this book is on the implications of genes for behavior that have already been detected in these early hours of a scientific revolution.

Ken Kendler, who laid foundations of our understanding of the inheritance of psychiatric diseases, has stated a set of principles which imply that we should not be pushing forward with naïve strategies or, more seriously, that we face ultimate obstacles in identifying genes for behavior. Prominent neuropsychiatrist Danny Weinberger once remarked to me that after listening to Ken's ten "thou shalt nots" he felt like an Israelite admonished for having worshipped the Golden Calf. The problems Ken has divined are real and at first might appear to form an intractably tangled knot. If a puzzle appears so complex as to be insoluble it can lead to inaction rather than having the more salutary effect of helping people to better design their studies. However, in science similarly complex problems have frequently been solved by swift cutting strokes and in many of these instances nothing would have been attempted if people had been cognizant of all the reasons why it should not work. As was once sardonically said, "ignorance is strength". Some experiments (e.g. unpowered human flight) will be attempted only by the wisest or most foolish. In genetics it has been proven over and again that gene discovery is a powerful tool to connect the most complex trait to an individual genetic determinant – the functional gene variant that drives the trait. In other words, genetic reductionism can work. This airplane can fly. If successful, the field of behavior genetics is not an arena of theoretical argumentation, but a practical exercise in gene discovery and the nature of the genes that are discovered. Throughout this book, and beginning with this chapter, I will be arguing from example, and I hope it is understandable, and forgiven in advance, that I have not filled this entire book with examples, which I believe are tangential to the theme of this book, of where the geneticist flew off course and the plane crashed or only flew a few feet before the geneticist ended up more or less where he had started.

In contrast to the often slow and incremental process of science and its frequent mishaps and sidesteps, the narrative of science is free to flash forward to such moments of discovery. Therefore, and except for the delay of this preamble, this chapter of the book introduces a gene that can cause people to commit murder. The rest of the book will be devoted to explaining how it can be true that there are genes such as this one "for" behavior, and how several have been discovered despite the formidable obstacles to which our attention has rightly been drawn. The theme of this book, and herein may lie the suspense factor, is that despite the discovery of genes that influence behavior, the locus of responsibility is preserved within the individual and not somewhere in the genes, or in environmental factors "beyond our control".

What happens if a gene for severe impulsivity is discovered, one that can cause people to commit murder, and is present in more than 100,000 people? Recently, my laboratory found such a gene: a functional knockout of *HTR2B*, a receptor for the neurotransmitter serotonin. As will be seen, this dramatic discovery had many precursors, including common genetic variants with more modest effects on molecular function and behavior. From the genetic association and *in vivo* (in the body) functional studies on those common variants, we anticipated that much inherited behavioral variation was also attributable to variants that are uncommon or population specific. Two examples of very rare variants that affect behavior and that are found only in individual patients and families will also be discussed. However, until 2009, it was usually impractical to find the rare variants that might exist in only one person or in their immediate family. Times have changed. The first human genome was sequenced at a cost of hundreds of millions of dollars. Today, and using technology for so-called massively parallel DNA sequencing, companies will sequence anyone's genome for between $5000 and $10,000 and very soon a genome will be sequenced for less than $1000. By massively parallel, I mean the ability to sequence millions or billions of DNA fragments simultaneously instead of sequencing one DNA fragment at a time or 100 at a time, as our machines allowed us to do only a few years ago.

The problem of reading our very long DNA "book" is overcome by cutting it into many small pieces that can be read very quickly, and at the same time. These hundreds of millions of little DNA sequences, which are usually only 35–100 DNA bases (DNA letters) in length, are then mapped back to their correct positions on the canonical map of the human genome, which is about three billion DNA bases in length. The sequencing is performed with high redundancy, with each DNA base being sequenced as many as 50–100 times so that the sequence variations that a person carries can be reliably detected. The technology is so ridiculously powerful that the typical student in my lab produces more data in their first few days than I did in years as a postdoctoral fellow.

For the relatively modest cost of sequencing his genome, what might an accused murderer interested in shifting locus of responsibility get? As we will see, potentially a lot. Potentially, his life. People have 25,000 genes, which provides ample opportunity to discover DNA findings that at least *look* powerful. For example, recent sequencing studies have revealed that the average person of European ancestry carries up to 100 "stop codons", but most are quite rare and their effects on disease risk are unknown. A stop codon can completely block the function of a gene, and in addition there are many other types of sequence variation that can be strongly predicted to alter molecular function and thus contribute to disease risk. In evaluating the effects of these gene variants we have to look beyond their strong effects on molecular function to their effects on behavior, which are usually much weaker and modified by other factors.

A COMMON GENE KNOCKOUT CAUSING SEVERE IMPULSIVITY

My laboratory applied the power of massively parallel sequencing to violent Finnish criminals at the University of Helsinki. When we studied these convicted felons, they were inpatients at a forensic psychiatry last headed by Matti Virkkunen, at whose retirement symposium I spoke last year. In Finland, many convicted violent offenders are assessed psychiatrically, and this work had yielded important findings about biological determinants of criminal behavior. For example, Matti and Markku Linnoila (more about Markku later) had found that elevated testosterone and low serotonin levels played important roles in impulsive and violent behaviors.

The prisoners studied by Matti are, as a group, not typical of convicted offenders in the USA. A larger number of them were impulsive violent offenders and arsonists who for the most part committed their crimes for no discernible purpose. Many had senselessly killed a friend, and often the murderer was intoxicated and committed the murder following some minor irritation. This type of behavior can be classified as intermittent explosive disorder (IED), except that in current diagnostic classifications the diagnosis of IED is excluded if in the context of alcoholism or antisocial personality disorder (ASPD). Almost all of these impulsive murderers were also alcoholic, and many had made significant suicide attempts or later completed suicide. With Matti and Markku Linnoila, we teamed up to conduct the first systematic genetic study of people with this type of extreme dyscontrol behavior. We also worked with an American psychiatrist, LaVonne Brown, who is a world expert in psychiatric assessment and impulsive behavior, and whose work on how to define impulsivity will be discussed later. Matti collected the DNA, extensive biochemical data, and psychiatric and behavioral histories from the criminals, their relatives and "normal control" men who lived in the Helsinki area, where a large fraction of the Finnish population is found.

Science really is too slow, and the "2B or not 2B" discovery was made only this past year, more than 15 years after Markku's life was taken by renal cancer. Matti and Markku are Finnish. It is frowned upon to characterize a people by their nationality, but for some reason it seems to work. Jaakko Lappalainen, who led some of the early work in my lab on genes and impulsivity, introduced me to the Finnish mentality with a story about a happy Finnish groom. Like the fact that Jaakko's hero was Axel Rose, that at first sounded like a self-contradiction, until it was properly explained. Then I was the only one who seemed to think it was funny. The evening before his wedding the groom was taken out drinking by his buddies and they all got stone drunk. Following the night of revelry and fellowship – certainly the last happy night of his life – his friends rolled him up in a carpet, as if he was Cleopatra, and delivered him to his bride's front doorstep, where in the morning she found him, dead of course. Very sad, but in a way funny. I laughed, but I'm still not sure where Jaakko, who seems so very happily married, stood on whether this was the best outcome for the Finnish groom. Jyrkki Vanakoski introduced me to the "Unknown Soldier". Another Finnish geneticist, Kaija Valkonen, was compelled to test the limits of new equipment by twisting the power dial further to the right than mechanically possible. A conflict ensued because cold-adapted Kaija left her bedroom windows wide open throughout the winter. The building's heating costs had skyrocketed, so the landlady wanted Kaija deported. Kaija is the type of woman who would be victorious in a wife-carrying race. Markku Koulu and Ulla-Marie Pesonen did great work in my lab and then greater work when they returned to the University of Turku and discovered a role for the neuropeptide Y gene in obesity. We will discuss neuropeptide Y in Chapter 16. It is a crucial mediator of stress resilience and its genetic variation plays an important role in anxiety and emotion, and via a gene by environment interaction with stress.

Clearly, gene by environment interaction has been at work over generations in Finland. Winter light, lichened granite, silent snow, cold waters and solemn forest were distilled into Sibelius' music and perhaps can in some way explain the sauna, if not how so many Finns came to possess the *HTR2B* stop codon. One Finnish man loved his wife so much that he almost told her. Another was ice-fishing with a friend. After a week of silence except for the wind, creaking ice and whispering of unknown forest-things, the man remarked, "It is very cold today." His companion replied, "Did we come to fish, or talk?" No one will really understand these Finns, least of all an American geneticist. Stranger still, when they decide to break the rules they break nearly all of them in unpredictable ways. Markku Linnoila was a child television star, motorcyclist and classical guitarist before he ever became a world-renowned biological psychiatrist. Such multitalented individuals can never be classified by nationality. We all know that nationality-based classifications don't work!

Hidden in the genome of some of the impulsive Finnish murderers we began studying in the late 1980s was a different Finnish mystery: an ancient stop codon mutation in the serotonin receptor, *HTR2B*, found by Laura Bevilacqua, an Italian psychiatrist. Helping train young postdoctoral fellows from all over the world is the best part of my job – and the saddest when they leave. I have to admit a special fondness for young neuroscientists from Italy: Chiara Mazzanti, Alessandro Rotondo, Francesca Ducci, Silvia Castrogiovanni, Nicoletta Galeotti, and last but not least Stefano Michelini the erstwhile soccer star, soccer coach, magnetic man and author of a strange little novel which I will advertise here: *Sauna*. This book, which I keep in my office as an inspiration or object lesson, describes a lesbian love affair and is apparently motivated by professional resentment of a paragon of Italian science. So far, so good; the lesson being, "Don't do that." On the other hand, *Sauna* contains page after page with the following information:

".."

".."

".."

When I first opened Stefano's book at a random place somewhere near the middle, I was fortunate and fascinated to see one of these "saunatas". I couldn't read Italian anyway, but here was something that resonated: pay attention to unstated thoughts and motives. Try to imagine the internal landscape. Then measure it with brain scanners and DNA sequencers. Otherwise, all too often all we really know is "..." and what our imaginations leave us. Another lesson seemed to be to write the way one wanted, and perhaps within the artifact that is a book in which someone might find a purpose, even if not the one intended.

The deep sequencing Laura performed and that led to the discovery of the HTR2B stop codon was entirely purposeful, and required the talents and intense efforts of many people with advanced skills and special ability to solve molecular and computational problems. This is "rocket science". Zhifeng Zhou, a molecular biologist with a wry sense of humor, worked directly on the physical part of generating millions of fragments of DNA sequence. Qiaoping Yuan, who earned his PhD in Forestry and became an expert in bioinformatics while assembling the rice genome sequence, devised ways of processing the masses of sequence data – more than 50 million sequence reads per run of the sequencing machine Zhifeng and Laura were using at that time. This sort of bioinformatics work would have been impossible, or taken a lifetime, without a network of high-speed computers linked together by Lisa Moore, our computer systems maven.

Genetics is an unusual domain of biology because from time to time an incredible complexity of data can be boiled down to a single reductionistic, explanatory result. In this case, the reductionistic finding was a stop codon that was actually causal for a very complex trait, severe impulsivity. The location of the stop codon in the 5HT2B serotonin receptor is shown in Figure 3.1. Each circle (or large dot) represents an amino acid building block of the receptor protein chain, which is folded so that it crosses the cell membrane seven times. When the stop codon is present this elegant architecture is guillotined after only 20 amino acids. Obviously, this small protein fragment is incapable of performing the normal functions of this serotonin receptor to bind serotonin at the cell membrane and initiate an intracellular signaling cascade.

What we can learn from the picture is that the HTR2B mutation is severe, equivalent to a "knockout" of the gene. Neurotransmitter receptors of this type have an elegant structure honed by millions of years of evolution and consisting of an extracellular portion (at the top), seven regions that span the cell membrane of the neuron (middle), and some intracellular regions that interact with signaling proteins (bottom). However, the stop codon terminates the protein after only a small 20-amino acid fragment is made, corresponding to only part of the extracellular domain and omitting the rest of the receptor.

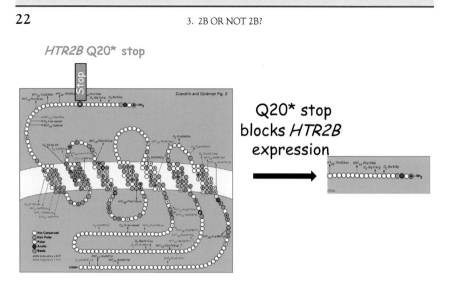

FIGURE 3.1   Location in the *HTR2B* serotonin receptor of a stop codon common in Finland. *(Source: Cravchik and Goldman)*

In males, the stop codon was more common in murderers, other violent offenders and alcoholics. Most of the male stop codon carriers who we had identified in the criminal population had also attempted suicide. The violent offenders with the stop codon had a characteristic pattern of behavior. Under ordinary conditions they were normal, but when intoxicated with alcohol they became highly irritable and dyscontrolled, engaging in violence that on a superficial level appeared senseless. For example, the criminal with the stop codon might have been angered and immediately tried to set fire to a nightclub full of people, or choked someone to death with his bare hands. As might be expected, the behavior of some of the murderers who carried the stop codon is as chilling as it is pointless. While drunk, one stop codon carrier was choking a woman who had irritated him; however, he then strangled the man who interrupted him. Convicted of murder and remorseful, he was imprisoned for several years, but following his release throttled to death another man. A remarkable result, which is probably due to a combination of causation, chance, and the unique genetic and social structure of Finland, is that the three double murderers in our sample were carriers of the *HTR2B* stop codon. In the whole country of Finland, double murderers are rare. Yet, here were three unrelated double murderers who all carried the same severe genetic variant. Counterbalancing this picture developed by studying severe violent offenders and comparing them to controls; we directly detected the stop codon in 174 Finns (and since that time our Finnish colleagues have found more than 1000), and learned that people carrying this genetic variant are cognitively within the normal range (although there may be nuances) and most are non-impulsive. By extrapolation, we know that the stop codon is found in over 100,000 Finns, but almost all are "normal". In some Finnish murderers the stop codon was a necessary factor, but it was not sufficient to account for the severely impulsive behavior of any of them.

In women, the effect of the *HTR2B* stop codon is different from men. Several were alcoholic but others were apparently normal, and none was known to have committed murder. Most interesting were the families, where the stop codon again was found in association with behavioral control problems in relatives, and again the exceptions were mainly women. Unlike the monoamine oxidase A (*MAOA*) stop codon mutation, which was found in only one Dutch family and will be discussed in Chapter 4, we were immediately able to find the *HTR2B* stop codon in a series of families. As mentioned, we extrapolate that more than 100,000 Finns, about 1 in 50, carry it. I believe that the high frequency of the stop codon, and its restriction to Finns, reflects the unique nature of the Finnish population, which is a population that has genetic and linguistic characteristics differentiating it from other European populations, and which has a number of disease-causing mutations that are either rare or completely absent in other populations. In fact, it was difficult to find the *HTR2B* stop codon in anyone whose DNA we did not collect in Finland. Laura Bevilacqua genotyped the DNA from several thousand other individuals from various populations distributed all around the world. Finally, she came to me in a state of excitement because she had found an American who carried the stop codon. Perhaps this was a variant that was not, after all, unique to Finns. We immediately contacted our collaborator, Emil Coccaro, at the University of Chicago to find out more about this person. She was an alcoholic, and her ancestry was … Finnish.

VALIDATING A HUMAN IMPULSIVITY GENE IN A MOUSE MODEL

An obvious problem with human behavioral genetics, and the human genetic study that enabled us to discover the *HTR2B* stop codon, is that scientists cannot readily perform controlled experiments. Human geneticists study the individuals, family constellations and populations that exist, and usually cannot manipulate environmental exposures. People who have the stop codon and who as a result are severely impulsive might for several reasons be underrepresented in epidemiological samples. This is one reason why it was important that we had available a sample of severely impulsive Finnish criminals and matched controls. The criminals could have declined to take part in our study, but few did. As we will discuss at more length, the Finnish men we studied choose their environments or have them thrust upon them, and some of their environments are very different. Also, if we are interested in the action of a gene such as *HTR2B* we have to be concerned with the effects of genetic background. What are the other 25,000 genes doing? On the other hand, what we care most deeply about may be the human behavior, and the genetic variant that alters behavior in the human might be rare or non-existent in an animal model.

Fortunately for the *HTR2B* stop codon story, we were able to evaluate the behavior of a relevant animal model, and thereby show that the discovery we had made in people had predictive validity in another species where we could control both environment and genetic background. Luc Maroteaux, a scientist who has long studied the *HTR2B* gene, had used recombinant DNA methods to target and disrupt the gene in mice. In these mice, other genetic and environmental factors are held constant, but the *HTR2B* gene has been knocked out. Key to the study of such an animal model is that impulsive behavior, novelty seeking and responses to novelty are measurable and correspond to human behavior. Another advantage is that we did not have to worry about underascertainment of mice that happened to be impulsive, because the mice, in contrast to people, could not wander off or decline to take part in our studies.

We will discuss the definition and measurement of impulsivity in people in more depth later. For now, it is sufficient to say that impulsivity of these mice was measured experimentally and in several ways, as illustrated by the "2B" and "not 2B" mice in the yin-yang clock face shown in Figure 3.2, credit for which goes to Luc and his lab. The patient

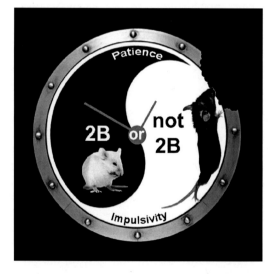

FIGURE 3.2   Impulsivity in mice, and the effect of a knockout of the *HTR2B* serotonin receptor, also missing in some people. *(Figure courtesy of Arnauld Belmer)*

"2B" mouse bides its time in exploring a new environment, even to seek food when hungry. It fears a novel object that is placed in its cage and takes its time to touch and explore this new thing that could be dangerous. Although hungry, the "2B" mouse cautiously lurks at the edges of a box biding its time before venturing to the center where it knows it is likely to find a food pellet, but where it is also more exposed to harm. In contrast, the impulsive "not 2B" mouse with its htr2b gene knocked out is less inhibited by fear, and intolerant to delay. When placed in a novel environment it rapidly explores it. When a novel object is placed in its environment it more readily touches it. Perhaps it is something tasty! If it is hungry it does not wait long before sallying into open spaces that mice fear and, at least this time, it finds a food pellet waiting. Later we will discuss that the "not 2B" mouse actually is different in how it discounts reward versus the amount of time it has to wait for a larger payoff. The "not 2B" mouse and the people who carry the *HTR2B* stop codon are not stupid but, as we were able to measure, the "not 2B" mouse is too impatient to wait a long time even though it has learned that the delayed reward would ultimately be greater.

What does the discovery of genetic variations with a role in impulsive behaviors mean, and what does it mean that we can even create a controlled model in the mouse that validates genetic effects observed in people? First, that the question of whether impulsivity and its resulting deleterious behaviors are in part inherited is settled. In retrospect, it could not be otherwise: behavior is the product of brain and the building blocks of the brain are subject to genetic variation. We are all neurochemically individual. Second, we've learned something about the nature of the inherited variations that alter impulsivity. With the genes discovered we can state that the genes that make some people more or less susceptible to behavioral dyscontrol work in different ways. Looking forward, we will see that some variants, such as Han Brunner's *MAOA* stop codon, are rare – the vulnerability genotype being found in only one Dutch family. Others, such as a different variant at *MAOA*, are common – someone in every gathering of people probably has the vulnerability genotype. Some, such as the *HTR2B* stop codon, are found in multiple families, but are restricted or relatively restricted to only one population. Several of the variants, and perhaps all, are strongly dependent in their action on contexts, for example being male, having a high testosterone level and having experienced an early-life stress. The genetic variant may be a necessary part of the behavioral syndrome, but alone it is not sufficient.

4

Stephen Mobley and His X-Chromosome

"If anyone slays a person, it would be as if he slew the whole people: and if any one saved a life, it would be as if he saved the life of the whole people."

مِنْ أَجْلِ ذَلِكَ آتَيْنا عَلَى بَنِي إِسْرائِيلَ أَنَّهُ مَنْ قَتَلَ نَفْسًا بِغَيْرِ نَفْسٍ أَوْ فَسادٍ فِي الْأَرْضِ فَكَأَنَّما قَتَلَ النَّاسَ جَمِيعًا وَمَنْ أَحْيا ما فَكَأَنَّما أَحْيَا النَّاسَ جَمِيعًا وَلَقَدْ جاءَتْهُمْ رُسُلُنا بِالْبَيِّناتِ ثُمَّ إِنَّ أَثِيرًا مِنْهُمْ بَعْدَ ذَلِكَ فِي الْأَرْضِ لَمُسْرِفُونَ (32)

Quran, Surah 5: Al-Mā'idah (The Table Spread) – سورة المائدة Verse (Aya) 32

We have just seen that genes that affect impulsivity and aggression are unlikely to work in isolation, because even the *HTR2B* stop codon does not have that type of unilateral effect on behavior. Despite some occasional over-enthusiastic reporting of gene effects on behavior and other complex traits, there is a general consensus that behavioral causation is multidimensional. However, it has occurred to defense attorneys that juries ought to hear about any factor that could predispose their client to commit a crime. It has also occurred to prosecuting attorneys that they should do everything possible to exclude that evidence. A courtroom is a laboratory where we can see what happens when forces collide, within the constraints of precedent and legality.

Stephen Mobley died when he was 39 years old, at 8 pm, on March 1, 2005, at a state prison located in Butts County, Georgia, south of Atlanta. He had received a lethal intravenous injection eight minutes earlier.

Minutes before that, he had been granted a stay of execution, but it was soon withdrawn. Mobley was a paradoxical figure and was made into a symbol even though he was not well suited to the role. In his final statement, he expressed thanks for the opportunity to atone for his sins. Then his executioners in an adjacent room injected Mobley with a combination of sodium pentothal, to put him to sleep; Pavulon, to stop his breathing; and potassium chloride, to stop his heart. Outside the prison, a small group protested, and vigils were held statewide. Afterwards, Laura Moye of Amnesty International, who was one of the few who attended the execution, said "We think the state should not be in the business of killing people to show that killing people is wrong." Friends and relatives of the victim did not witness Mobley's death, and were not reported to have held vigils. However, the victim's family apparently supported commutation of Mobley's penalty to a life sentence.

Mobley had committed murder 14 years earlier while in the midst of a three-week spree of armed robberies of restaurants and dry cleaners. While robbing a Domino's pizza restaurant about an hour's drive northeast of Atlanta, he shot the night manager in the head. This man, who Mobley found alone in the restaurant, was a 24-year-old college student named John Collins. A few weeks later Mobley was caught following a high-speed chase through Atlanta after another robbery. He was still armed with the gun he had used to kill Collins. Mobley is said to have confessed to a prison guard that he ordered Collins to his knees, made him beg for his life and, after his victim began crying, killed him execution style with a shot to the back of the head. As described by a journalist, Mobley was a funny, charming and affable man who related well to people, even in the minutes prior to his execution.

THE KALLIKAK EFFECT

Mobley was born into a prosperous family, and was represented by a former Attorney General of Georgia and a former county District Attorney. These attorneys pointed to the long history of criminal behavior in the Mobley family, claiming that his genes made it inevitable that he would do the same. The behavior of the Mobley family allegedly resembled the bad side of the legendary Kallikak family, whose pedigree was featured in eugenically-tilted genetic textbooks of the first half of the twentieth century.

The Kallikak family tree had two branches because the founding father was a Revolutionary War soldier who dallied with a barmaid but later married a Puritan girl. Supposedly, the descendants of the barmaid were uniformly morally degenerate and cognitively defective drunkards and thieves. This side of the family tree was depicted with dark twisted limbs and toadstools at the roots. On the other hand, the descendants of the Puritan wife were chaste and upstanding. That side of the family tree had smooth limbs and flowers in bloom at the base. Was Mobley's Kallikak-like family history relevant as a mitigating factor? As law professor Debbie Denno pointed out in her analysis of the Mobley case, being a regular churchgoer or loved by one's family may be admissible as evidence. However, it has also been reported that Mobley had "Domino" tattooed to his back and hung a Domino pizza box in his prison cell. So should inherited predisposition be off the table, whether as a mitigating factor or perhaps even to make the case for guilt and irredeemability? In a case where genetic testing for Arizona death row inmate Jeffrey Landrigan, would-be assassin of Congresswoman Giffords, was rejected, a lower court judge wrote, "The potential for future dangerousness….inherent in Landrigan's alleged genetic predisposition to violence would have negated its mitigating capacity for evoking compassion." Also, an Idaho judge accepted evidence for another defendant's "propensity to commit murdermurder," and used it to help justify the death sentence.

MOBLEY DEMANDS A GENETIC TEST

Mobley's attorneys petitioned that Mobley be tested – at state expense – for a predictive variant in the monoamine oxidase A (*MAOA*) gene. Like the 5-hydroxytryptamine (serotonin) receptor 2B (*HTR2B*) stop codon we discovered in Finns, which is found in over 100,000 of them, the *MAOA* stop codon is a severe genetic variation which blocks the function of a key protein involved in neurotransmitter function. It had recently been discovered by Han Brunner, in one Dutch family. Unlike the *HTR2B* stop codon, this *MAOA* stop codon was never found outside this one family. Various geneticists, including me, were contacted by defense attorneys to test their clients for the stop codon in *MAOA* (I declined). Meanwhile, my lab and others were investigating other impulsive individuals and families, including some of the same ones in whom we eventually discovered the *HTR2B* stop codon, for the presence of Brunner's genetic variant, but not finding it.

The judge rejected having Mobley's *MAOA* genotyping paid for by the state, or accepted as evidence. This decision made good genetic sense. The *MAOA* gene is on the X-chromosome, which an XY male inherits from his mother, and the male has a 50 percent chance of inheriting the stop codon variant from his mother if she is carrying one copy of the variant. The *MAOA* stop codon variant appears to be highly penetrant: the males in the Dutch family who inherited the variant had a behavioral dyscontrol (impulsivity) syndrome. This leads to an X-linked recessive pattern of transmission in the family: carrier mothers,

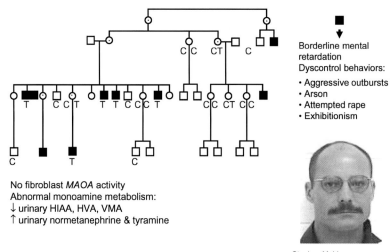

FIGURE 4.1   Brunner syndrome: X-linked dyscontrol in a Dutch family due to the monoamine oxidase A (*MAOA*) C936T stop codon (Brunner et al., *Science*, 1993; Brunner et al., *Am. J. Hum. Genet*, 1993).

50 percent of male offspring affected, other affected males on the maternal side of the pedigree. Instead of this pattern, Stephen Mobley's family had multigenerational impulsivity with transmission from father to son and son to grandson, as well as violent females (Figure 4.1). An X-linked stop codon was therefore one of the least likely causes of his behavior.

COMBINING GENE AND HORMONE TO PREDICT IMPULSIVITY

The story of functional genetic variants at *MAOA* did not end with Brunner's rare stop codon. Because *MAOA* will probably follow the same pattern of variation as other genes, it could have been predicted that eventually hundreds or even thousands of variants will eventually be found at this same gene. Of course, most of these will be rare. Several years ago a common functional polymorphism was discovered in the same *MAOA* gene that Brunner studied. The alteration in function was discovered by Dean Hamer, who will appear again later (in Chapter 15) as the discoverer of the "gay gene". This new *MAOA* variant altered the expression of *MAOA* at the regulatory level: at the level of DNA transcription, the process by which the *MAOA* gene is expressed as an RNA transcript that will be translated into the enzyme protein. The polymorphism is of the short tandem repeat (STR) type involving varying numbers of copies of a short DNA sequence found consecutively, or as in the name, in tandem. It has several common variants that differ in length owing to the number of repeats that are present. At least two of these length variants are associated with a lower level of *MAOA* expression and *MAOA* enzyme activity, and it is these relatively common lower activity *MAOA* variants that have been associated with impulsive behavior.

CARRYING KOHL TO ITALY

The Court of Appeals in Trieste recently reduced a convicted murderer's sentence based upon this *MAOA* genotype and some other genetic and brain imaging evidence of a type we will discuss later. Abdelmayout Bayout, a citizen of Algeria, admitted stabbing and killing Walter Felipe Novoa Perez in 2007 because the victim had insulted Bayout over his kohl eye make-up. Although Bayout had already had his sentence reduced based on psychiatric illness, the apellate judge, Pier Valerio Reinotti, requested an independent evaluation. Two neuroscientists at the University of Padova evaluated Bayout and informed the judge that Mr. Bayout's genotypes at *MAOA* and four other genes made him prone to violence following provocation. In his decision, Reinotti stated that he was strongly compelled by the *MAOA* genotype evidence, and reduced Bayout's sentence by an additional year, to a total of eight years and two months. According to the judge, Mr. Bayout's genes "would make him particularly aggressive in stressful situations". As has been discussed, the genotypes entered into evidence would have had little predictive value. Furthermore, based on the earlier examples from Montana and Arizona, it is equally possible that another judge could have come to exactly the same conclusion about the effect of *MAOA* on Mr. Bayout's behavior, but therefore increased the sentence.

THE STATE OF DNA IN PREDICTION OF VIOLENCE

The power of *MAOA* genotype, *HTR2B* genotype or other genotypes to predict violence of an individual is extremely low. However, the number of cases in which DNA evidence has been introduced in an attempt to demonstrate that a defendant was genetically predisposed to violence, depression or addiction is rapidly accelerating. According to Nita Farahany, a bioethicist at Vanderbilt, there have been 200 attempts to introduce mitigating DNA evidence in cases tried in the USA during the past five years and a few have been successful at the penalty phase of these trials. For many reasons that are a main preoccupation of this book, there is strong resistance to the use of such evidence, but as Bayout's lawyer said, "My client is clearly an ill person and everything that allows the Judge to better evaluate the case and to decide the right sentence should be investigated."

Therefore, it is worthwhile to discuss just how this common *MAOA* variant might be most accurately used in individual cases where behavioral prediction and explanation is so difficult, as contrasted with their use in large studies identifying group differences. If one genotype is not highly predictive on its own, why not combine it with other information? To identify circumstances where *MAOA* genotype could help predict or understand one person's behavior we should understand the circumstances where group behavioral predictions are powerful. The *MAOA* variants have been repeatedly associated with impulsivity, but not in a general context. The first indications of behavioral prediction with *MAOA* have emerged only when the effect of the gene was measured in combination with other powerful predictors of behavioral problems. For impulsivity and aggression, two powerful predictors are testosterone and trauma, and we are still early in the process of understanding how such combined information can predict the behavior of an individual, as opposed to a group of individuals.

The relationship of impulsivity and aggression to testosterone is obvious. Worldwide, and across cultures, men are far more likely than women to commit violent mayhem. The ratio of males to females on death row approaches 100 to 1. As a predictor of impulsivity and violence, the Y-chromosome is an informative genetic marker, or to put it a better way, the absence of a Y-chromosome is compelling. Women are statistically unlikely to commit violent crimes and when they do it is more likely that there was a specific motive, a psychosis or a man involved. In most non-human primates, including our closest relatives such as chimpanzees, gorillas and orangutans but also numerous Old World monkey species, males are also more aggressive. It appears that one important reason is the much higher testosterone levels of males. As can be seen on the left side of Figure 4.2, higher testosterone in men correlates with increased aggression, as measured by a lifetime aggression score called the Brown–Goodwin scale, which we will discuss in Chapter 6. It should also be mentioned that one reason the relationship between testosterone and aggressive behavior is so strong in this figure is that the sample included highly impulsive and aggressive men. It is difficult to find a relationship between a hormone and an outcome if the sample does not include very many individuals with the outcome.

The effect of stress interacting with *MAOA* and other genes to lead to impulsive behavior is less obvious, but very powerful. As seen above, Francesca Ducci, Rickard Sjoberg and I discovered that in Matti Virkunnen's impulsive criminal offenders, higher testosterone levels

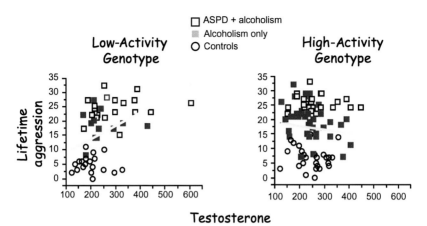

FIGURE 4.2 Monoamine oxidase A (*MAOA*) genotype and testosterone interactively predict aggression. All subjects were male. Lifetime aggression was measured on the Brown–Goodwin scale; testosterone concentration was measured in cerebrospinal fluid (pg/ml). ASPD: antisocial personality disorder.

indeed predict higher levels of aggression, but this happens only in the men who have the common lower activity *MAOA* variant that is itself associated with aggression. The combination of high testosterone level and low *MAOA* activity genotype represents a "double whammy".

This gene by endocrine interaction is so far one of the few of its kind, but as will be seen the discovery of interactions of genes with other factors, especially stress, represents one of the most salient accomplishments of genetics of complex behavior. Gene by stress interactions are the topic of Chapter 16. Where will gene by sex, gene by endocrine and gene by stress interactions lead us? Fast-forwarding to the present, a defense attorney defending a murder case might make much better use of the *MAOA* gene, but he would have to work a little harder at it, and more than genotype might have to be measured. Seldom will it be sufficient to consider one factor. Finally, some diseases, but most especially most of the severe socially defined and social-context influenced behaviors, are likely to not only be multifactorial in nature but also dependent on overall societal context.

Dial Multifactorial for Murder
The Intersection of Genes and Culture

"Good and evil we know in the field of this world grow up together almost inseparably...."

Areopagitica, John Milton

One way of introducing chapters is with a quotation and an explanation of the possible relevance. On the other hand, National Public Radio accompanies its features with little musical interludes that suggest a mood. Another way to engage the reader is to begin a chapter with a story or parable. Instead of explaining quotations and in general using thousands of words to express what could be said with hundreds, perhaps they should be more self-revelatory, at the expense of adding self-referential paragraphs such as this one, to at least give the reader some better idea of what has gone on "behind the curtain".

I have chosen to begin this chapter on the complex origins of murder and violent behavior with a particularly brutal and horrible murder story. Who is not at once repulsed and attracted by a diabolical act? However, there are many murders to choose between, providing the opportunity to select a case to prove any theory of human behavior. Perhaps the murderer was a hardened criminal who was unwisely paroled, perhaps he was an impulsive murderer with an unknown stop codon who then became a double murderer, or a victim of child abuse, or uneducated, or impoverished, or a racist, or a gun nut, or a religious fanatic or of the wrong political orientation.

As a scientist I like to think that the plural of someone else's anecdote is my data. However, the choice of the "anecdata": the vignette, the quotation or the mood music, sets the stage, and often overwhelms the impact of reasoning based on larger trends, statistics or systematic thinking. With regard to murder, Mark Kleiman wrote, in *When Brute Force Fails: How to Have Less Crime and Less Punishment*, that we go forward on the basis of "big cases" to make "bad laws". We are haunted by the face of a child who should not have died and who should not have died in some horrible way. That's why laws are named after real victims, for example "Megan's Law" and the "Brady Act".

A MURDER IN THE LAB

However, I do have a personal reason to write about one particular double murder, because the murderer took the life of a close colleague, Dr. Michelle Filling-Katz. I hope that the telling of her story is taken as respectful, non-gratuitous and non-exploitative of her memory. I hope this personal drama helps the discussion of biological origins of murder and is one more legacy of the victims, Michelle and her husband Norman.

Michelle, a dynamically talented pediatric neurologist, was a redheaded lady with an outspoken approach to science. If anyone dragged their heels or put an obstacle in the way of her care for patients or her research she let them know. Michelle herself endured the ravages of autoimmune disease, and this gave her a special insight into the suffering of her patients, many of whom had von Hippel–Lindau disease, a genetic disease whose manifestations include cancer of the kidney. She was only 36 years old, and had joined my lab two years before her murder. Michelle and Norman Katz had both been army physicians. Colonel Norman Katz was a Green Beret who had served with the 82nd Airborne Division in Vietnam. He received two Bronze Stars for bravery, and retired only in 1989 after having been Chief of Pediatric Ophthalmology at Walter Reed Army Medical Center.

Michelle and Norman were shot dead in their kitchen on August 11, 1992. The murderer was Michelle's stepson, Jayant Katz, a 20-year-old architecture student who I had briefly met a week or two before the murders. The only thing I noticed about his behavior was that he was very concerned about his car. Jayant Katz had schizophrenia, a disease which more frequently leads people to impulsively harm themselves, but as we will discuss in Chapter 7, sometimes leads people to harm others, even ones closest to them.

What do we learn from such murders? Would Michelle and Norman be alive today if Jayant Katz had not had access to a gun? Or should we take away lessons in family dynamics? Perhaps Jayant Katz was exposed to violent cultural influences. Perhaps the schools were at fault. Or should we understand that Jayant Katz's schizophrenia – a heritable disease – played the most important explanatory role? Surely his psychiatric illness played some major role. In the Maryland Circuit Court located in the county where I live, Jayant Katz pled guilty to two counts of second degree murder. He said to the judge: "Yes. I killed my mom – it's my stepmom actually – and my dad. Bang. The parent dead. A thousand voices fill my head."

MISSING PUZZLE PIECES, AN OBSTACLE TO BEHAVIORAL REDUCTIONISM

Psychiatric diseases can play a role in violence, and genes can do so either by contributing to such diseases or through their independent effects that cut across diagnoses or co-contribute to various psychiatric diagnoses. Thus far, we have shown that functional variants of genes, monoamine oxidase A (*MAOA*) and 5-hydroxytryptamine (serotonin) receptor 2B (*HTR2B*), can predict violence. The Y-chromosome predicts violence. The interaction of testosterone with *MAOA* genotype predicts violence. Such particulate genetic findings can be useful for understanding individual behavioral differences and predisposition, but they do not constitute a reductionistic understanding of the problem of violence in societies. Murder is multifactorial. Even the most lethal genes are not highly predictive. Although they might be predictive if we fully understood the contexts in which they lead to violence, it is fair to say that we have only begun the process of understanding the interactions. It is unsurprising that we do not understand these interactions because no progress could be made until the genes were isolated.

To put it another way, we might routinely expect not to very well understand the role of environment in violence because up until very recently we had failed to identify any of the genes with which environment interacts. Scientists with a primarily environmental perspective on the origins of behavior often demand of geneticists that they measure the environment. If only we knew how. Increasingly, geneticists are being shown how, and increasingly geneticists are showing scientists performing longitudinal and family studies how to incorporate genotype into their work. Also, and as will be seen in Chapter 16, where we take up gene by environment interaction studies, it is often not enough to measure the environment. One has to study people who have experienced the relevant environmental exposures.

The overall problem for solving complex jigsaw puzzles of behavioral causation is that it is very much more difficult without all the pieces in hand. As a schizophrenic, Jayant Katz killed his parents, but few schizophrenic boys commit murder. Three of our *HTR2B* stop codon carriers became double murderers but most live their lives with no indication that this genetic variation is present. Clearly, there are some pieces of the puzzle that elude our science. On several occasions, my daughter Evir (who is now a teacher) instructed our family in that simple lesson by holding back a piece from the 1000-piece puzzles we liked to assemble over winter break. As the pile of pieces shrank, her brothers, Aaron and Ariel, would begin to suspect that there was again a puzzle piece missing, and would know full-well who was the culprit. They would shout as children do: "Mommy, Daddy, our bad sister Evir has done it again! She's hidden a piece!" But no matter what methods of interrogation were applied Evir never would admit to the crime. The children would then do what scientists have to do, which is solve the puzzle with a missing piece. At the last moment she would triumphantly pop in the last piece – which was usually something rather essential, like the fluke of the white whale or the eye of God. Had Evir become a scientist she probably would have been renowned because we often better remember those who place the capstone.

Fortunately, scientists are clever people who enjoy a challenge. Now that we have a few of the genetic pieces of what is a really complicated multidimensional puzzle of behavior it does appear that some progress is finally being made. Nevertheless, at this point, and using genotypes only as isolated predictors, the most informative genotypes are simply not very predictive. As discussed earlier, the *HTR2B* stop codon is found in a large fraction of Finnish murderers, especially ones convicted of double homicide. However, it has a rather high allele frequency: greater than 1 percent and about 1 in 50 Finns carry this variant. As is worth saying again, we know that almost all of them are normal, and very few will ever murder anyone.

WHY ARE SOME SOCIETIES MORE VIOLENT THAN OTHERS?

Just as we can understand the effects of genes without claiming to have found the gene predictor, we also do not have to wait for the full decoding of gene by environment interactions to find effects of environments that predict the behaviors of groups of people. Like the predictive value of individual genes, the predictive value of individual environmental factors is expected to be slight, and it is. Why does one society have more murder and violent crime than another? Why, within societies, do murder rates substantially decrease, and occasionally increase, across generations? Genotype can play only a minor or non-existent role in determining these transnational and secular variations. Although allele frequencies have not changed, poorly understood social changes can double or halve the murder rate. Countries whose peoples are of very similar genetic makeup have dramatically different murder rates.

Volumes are written on the subject of the origins of violence, but they usually seem directed towards proving some theory or another and underestimate the role of biological factors such as the importance of being male, young, high in testosterone and psychiatrically ill, and they do not consider the problem of genotype. For murder in particular, it is also important to come to terms with the fact that the event is rare. The USA has one of the world's highest murder rates, but the rate is only about 5 per 100,000. However, all of the predictors of murder are much more common than in many other countries. For example, guns are a predictor of murder but there are more than enough guns in the USA to arm every man, woman, child, infant and household pet. Yet few of these guns will ever be used to murder anyone. The conversation about the origins of murder should be directed towards understanding the behavior of that small percentage of society that commits those acts, and to understand that behavior it will undoubtedly be necessary to piece together multiple pieces of a causal puzzle that differs from one murderer to the next. This is why oversimplifications such as "guns cause murder" are as toxic as the overexaggeration of the role of some genetic variant. However, it also illustrates why the genotype is important even though it will only rarely lead to the event. None of the predictors will act that reliably and yet we should want to identify all the predictors and their interactions.

Decades of careful scholarship by criminologists and historians on the environmental origins of murder have, unfortunately, not gotten us very far towards understanding those origins. There is no agreement whatsoever on the major determinants, and the theories are quite diverse and sometimes seem to tell us more about the scholar than their subject of study. In *A History of Murder: Personal Violence in Europe from the Middle Ages to the Present*, Pieter Spierenburg, a professor of historical criminology, hypothesizes that murder rates are inversely proportional to the progression of the "civilizing process", the whole class of behaviors requiring self-control and the increasing ability of modern states to monopolize the use of force, disarm citizens and enforce order. The "uncivilized" or backward nature of the USA would thus explain its two-fold higher murder rate compared to Western Europe. It isn't just the guns, it's our culture. Randolph Roth, in *American Homicide*, takes up a thread developed by Eric Monkkonen and concludes that the high rate of murder in the USA is due to our political institutions and public faith in them, trust in the integrity of public officials, social solidarity based on race, religion and political affiliation, and faith in the social hierarchy as an avenue to achieve respect. Democracy requires dissent, and in Roth's view when people dissent from a president whose politics they don't approve of, this leads to an increase in the murder rate. The message seems to be to elect "good" presidents, but we can always be sure that at least 30 percent of Americans will think the President is terrible. Historian–criminologists reach some convenient destinations (Americans should be more civilized, Americans should elect good presidents, Americans should stop clinging to their guns). However, to make sense of this rare behavior it seems better to identify the factors and rank them by hierarchical importance. Factors that will ultimately be integrated with predisposing genotypes include economic disparity and male sexual competition, youth, guns, poor medical care, poverty, drugs and mental illness.

GUNS OR PEOPLE?

Each one of the individual environmental determinants just listed is itself a complex story, as illustrated by guns. A gun can be found in almost one of two American homes, and on an annual basis approximately 10,000 of the 15,000 American murders are committed with guns. Does this mean that guns cause murder? Not really. Murder rates were never higher after guns were invented than they were before guns existed. In medieval Europe the murder rate was about 35 per 100,000 and even by 1500 it was still about 20 per 100,000. At present, and after the invention of the gun, the murder rate in Western Europe is less than 2 per 100,000, and has been for the past century. Several countries with low murder rates are among those with the highest gun ownerships. In Switzerland, young men between the ages of 21 and 32 train for a few weeks a year and receive M-57 assault rifles and ammunition, which they are required to keep in their homes; once the citizen is discharged the M-57 is replaced with a bolt action rifle. In that country of seven million people there are at least two million publicly owned firearms – more than one per household. However, violent crime is rare in Switzerland – so rare that politicians usually don't have police protection. Whenever the Swiss example is mentioned, a series of differences between Switzerland and other countries is then raised by advocates of gun control: wealth, tradition, culture, lack of problems with drugs, lack of other social problems; however, that's exactly the point. Switzerland is not unique in having high gun ownership and low murder rates – other examples are Israel and Finland, and in fact the USA is the exception that, for gun control advocates, "proves the rule".

What about kinship and sexual competition? Anthropologist Claude Levi-Strauss studied the Yanomama, a remarkable and threatened people who live in regions of what is today northern Brazil and southern Venezuela. Distilling his incredible life work on the Yanomama into a few sentences, these so-called "Fierce People" had extraordinarily high rates of murder and frequently there were violent struggles between villages. From a geneticist's perspective one of the most interesting aspects of the intravillage and intervillage violence was that it was closely associated with sexual competition between males: those with the most kills had the most wives and children and struggles between villages often involved raids to capture women. Also, the likelihood of conflict between individuals and between villages decreased with higher kinship (genetic relationship). As villages increased in population these factors would tend to lead to fission, even though a larger village might have been better able to fend off the attacks of neighboring villages. Within the village there were several levels of conflict resolution. The first stage involved one belligerent hitting the other up the side of the head and the other one reciprocating. In the second stage the flat side of an axe was used. Next was the "pole stage" in which a pole was brought down upon the head of the opponent (the obvious question being, "Who gets to go first?"). The next stage, which was really more an example of conflict than conflict resolution, involved poisoned arrows. However, at least they did not have guns.

Personally, and I hope to the reassurance of some who would otherwise immediately label me "uncivilized", I favor some gun control measures. However, this is not because there is any good evidence that murder rates and crime rates are decreased by gun control. In some jurisdictions that instituted gun bans, homicide rates rose dramatically; for example, they doubled in Washington, DC from 1976 to 1991 following a virtual ban on handguns. The ban was ineffective, but clearly some factors other than guns were responsible for the large increase. In India, guns are not widely available but according to Indian police authorities at least 2500 brides are burned alive every year, and it was reported by *Time* magazine that in 1995 the number of deaths by wife burning was 5800. The usual procedure is that in-laws, angered because of lack of fulfillment of a dowry or demands for additional dowry, set the victim afire using kerosene intended for cooking stoves, or gasoline or whatever other flammable liquid is at hand. Perhaps vindictive relatives would be less likely to burn these defenseless young women if a few of them unexpectedly turned out to have a gun. In a 1982 survey of imprisoned US criminals, one-third actually admitted to being "scared off, shot at, wounded, or captured by an armed victim".

Despite its high rate of gun ownership and "gun culture", the USA does not lead the world in murder. Because many murders such as wife burnings are not recorded as murders, the murder rate in India (and other countries) is probably underestimated but is probably still less than that of the USA. However, the non-gun murder rates of Taiwan, the Philippines, South Africa, Mexico, Brazil, Estonia and Columbia exceed the total US murder rate, and the total murder rate of at least six countries exceeds that of the USA by greater than three-fold. Other than immolation and shooting, people find many alternative methods for killing. In the data sets I have collected and in the trials in which I testified, a variety of methods was used to commit murder. If a gun was not available, and sometimes when it was, another weapon at hand was often used. Physicians know that guns cause horrific wounds, and in some ammunition is designed to do so, for example by piercing protective armor or by expanding in the body, as hollow-point bullets do. I've treated those wounds, in survivors of course. However, as the burned brides could attest, if only they were still alive, the human body is fragile and easily damaged by bludgeoning, burning, stabbing, poisoning, being crushed by a car, poisoned arrows or being pushed from high places. Also, an interacting factor with means of injury is *medical care*, which has had a major effect on the murder rate. Today, many assault victims survive who only a few generations ago would have died and been counted as murder victims. Frequently, those survivors who do not appear in the murder statistics, have suffered devastating losses in quality of life, productivity and longevity.

CAN GUN CONTROL CIVILIZE PEOPLE, AND IF SO, WHAT DOES THAT MEAN?

Closing the loop on Spierenberg's civilizing influence and social control argument, there is a flip side to the sense of individual empowerment provided by gun ownership. The individual who has the mindset that she (or he) can defend herself may be more likely to act out on their impulses, perhaps with a gun or perhaps using whatever comes to hand, rather than seek the help of the system. One can judge for oneself whether that is a good thing. Some people would like a society where there is more social control and less potential for an individual to disrupt things. Others, of the "Ayn Rand school", "Live Free or Die license plate slogan school", "Eat my dust school of driving school" or "How's my driving? Call 1-800-EATSHIT", believe that the occasional death by shooting – and even accepting that these are usually tragic accidents and suicides – is a worthwhile price to pay for individual empowerment. Reasonable people, including Supreme Court justices, disagree over whether the Second Amendment to the US constitution guarantees the right of individual citizens to own guns. So as not to turn a book about genetics and society into a legal brief, and turn off many readers, I will simply state the obvious, which is that the Second Amendment was written as part of the so-called Bill of Rights for the common man, and not written for the purpose of preserving the rights of militias (except that in more archaic American usage militia was a word for a group of citizens). The sea-change was the extension of the Bill of Rights to women and former slaves, and to some extent, to corporations (groups of people). In the future, will guns be regulated, perhaps by constitutional amendment? Or perhaps people will change such that even with a weapon at hand or behind the wheel of some powerful vehicle they will be less of a threat to each other? If the schizophrenic stepson of my colleague Michelle Filling-Katz had not had access to a gun (obtained from his biological mother) then Michelle and her husband might have been alive today. Can such anecdotes, powerful as they are, justify policy?

VIOLENT YOUTH

A strong factor that drove murder rates upwards was the baby boom after World War II and their second generation offspring. With the aging of these cohorts, and the incarceration or death of some of the most violent, murder rates declined. This phenomenon was strongly observed in Washington, DC, where I live nearby. This change in demography dramatically reduced the number of hot-headed male perpetrators and potential victims, so it is not surprising that the murder rate has declined. Police chiefs and politicians claim credit for that decrease, but they do not deserve that credit any more than the previous ones deserved the blame for the original epidemic. The same error is usually made in comparing transnational rates. Murder rates in countries with low birth rates and an age distribution weighted to more mature individuals, for example Western Europe and Japan, will naturally be lower. Also, these age effects are probably not merely arithmetically proportional. Over longer periods the age structure of a population is likely to affect personal expectations, leading to a culture in which violence becomes more normative, as in the Yanomama. It is this complex arena in which genes altering propensity act.

Distorted Capacity I
The Measure of the Impaired Will

"Men at some time are masters of their fates;
The fault, dear Brutus, is not in our stars,
But in ourselves..."

Julius Caesar, Act 1, scene 2, William Shakespeare

All people make decisions, but some are inherently more patient and reflective, while others, as we learned in the "2B or not 2B" story, make

Our Genes, Our Choices
DOI: 10.1016/B978-0-12-396952-1.00006-3

decisions that are relatively impulsive. In each case a person has made a choice to engage in a conscious or *explicit* behavior. People also perform many actions *implicitly*, which is to say without conscious thought. Such actions can also be termed unconscious, but not subconscious, an ill-defined term.

Breathing is an example of the way people ordinarily transition between unconscious and conscious behavior. We ordinarily do not think about when to take a breath, and certainly not while sleeping. The behavior is unconscious. However, we can at will initiate or inhibit breathing consciously, and often do when we are instructed to do so by a doctor, practice yoga or breathing exercises, prepare to dive under water, or hold our breath while beneath the surface. Everyone has experienced that the urge to breathe becomes progressively stronger, and suppressing a breath requires increasingly strong conscious control, until loss of consciousness and reflexive breathing. Similarly, a person may unconsciously withdraw a hand from hot metal or from icy cold water. However, people have different degrees of capacity to consciously hold their hand against the pain, and can weigh the potential consequences.

Impulsivity, which at its extreme is action without forethought, an "implicitization" of behavior, is an important, normal behavioral trait. Hans Eysenck, the legendary psychologist who factor-analyzed human personality, pointed out that humans need impulsivity to enable them to initiate behavior. Following in the path founded by Eysenck, Cattell and others, psychologists can measure personality and temperament relatively easily using questionnaires. Although there are many such questionnaires and personality scales it is a remarkable fact that some common personality factors always emerge when the results are analyzed using a statistical methodology called factor analysis. Factor analysis takes the responses and asks, in a hypothesis-free fashion, whether there are response patterns that cluster together. Of course, whether a factor can be discovered depends to some extent on the questions asked. So, for example, the person who says that they are "afraid of spiders" is also more likely to report that they are uncomfortable speaking before large groups, leading to the identification of a factor associated with anxiety and introversion. One factor that emerges time and again even when the questions are varied is a trait associated with extroversion, novelty seeking and diminished behavioral restraint, or in other words impulsivity.

THE INHERITANCE OF IMPULSIVITY, AND WHAT IT MEANS

In twin studies, including studies of twins raised apart, the inheritance of such personality traits has consistently been found to account for 40–60 percent of the observed variance in the trait. As is discussed elsewhere in the book, the basis of the twin method is to compare the resemblance of identical twins to non-identical twins, and for personality traits identical twins tend have a much stronger resemblance. This is quite remarkable because of the crudity of the measurement tool and the multitude of non-genetic factors that could alter personality or the way that individuals respond to questionnaires of this nature. There are at least two important lessons to be drawn from the heritability of personality. The first is that these simple questionnaires do a remarkably good job at measuring *something*. We know this because heritability will always be limited by the precision of measurement. For example, if the precision of measurement is 70 percent, the maximum heritability would be 50 percent (0.7^2), which is about what the heritability of these personality factors has been observed to be. Assuming that environment also plays a role in personality, the precision of measurement of these personality factors is probably higher than 70 percent. The lesson to be drawn from the heritability of personality traits is that there are genetic variants that influence personality that are inherited from parent to child, and whose sharing between siblings makes them more similar. The third point is more subtle, and is based on the fact that even siblings and fraternal twins who share only 50 percent of their genes resemble each other in personality far more strongly than random pairs of individuals. This is important in understanding how the genes work. As will be described in full later, one possibility for gene action on behavior is that only particular combinations of genetic variants, acting in concert, can produce a given type of personality. The other possibility is that the action of gene variants one carries tends to be additive, and that the effect of a particular genetic variant, such as the 5-hydroxytryptamine (serotonin) receptor 2B (*HTR2B*) or monoamine oxidase A (*MAOA*) stop codons, can be understood without necessarily understanding everything else. This conclusion is frequently disputed among behavioral geneticists, but since no one is really arguing about the data showing strong resemblance between siblings and fraternal twins, all we can say is that whatever our philosophy of genetics this seems to be the way the genes are actually working.

The additivity of gene effects in behavior is fortunate for the process of gene discovery, because statistical analyses allowing for interactions of genetic markers are inherently less powerful and more difficult to interpret. When many tests are performed, one has to correct for the possibility of a lucky guess. Otherwise science becomes like a horse race in which the bettor who buys tickets for 100 combinations is more likely to hit the Trifecta for Win, Place and Show, independent of skill at judging horses. We have to be careful to avoid the "winner's curse", which is the phenomenon that if many people play a guessing game someone has to win, and the lucky winner can too easily delude himself that he "has a system". With a million genetic markers, about 500 billion two-way interactions can be tested (1 million/2). Any two-gene interaction discovered by such blanket testing would require very strong statistical support to confidently state that the finding was real and did not happen merely because the geneticist has bought 500 billion lottery tickets. The penalty to be paid for the blunt-force approach of testing all the interactions is larger, deeper and more expensive genetic studies. However, once having found genes "for" behavior, such as *HTR2B* and *MAOA*, we can then begin the more difficult process of determining how they interact, but at least the number of two-way tests is limited, because so far only a few genes are known to alter behavior. Even with a small number of genes, some will wish to test for higher-order interactions (three-way, four-way), again expanding the number of tests. If you want to win the lottery, you have to buy a ticket.

IMPULSIVITY DIFFERS FROM PERSON TO PERSON AND FROM SPECIES TO SPECIES: THE GORILLA AND THE KING

The heritability of impulsivity has a very important limitation that is based on the fact that what is on average true for the group is often false as applied to the individual. Impulsivity and its related personality characteristics such as novelty seeking are descriptors that are in some very obvious ways and to greater and lesser extents socially bounded. Whatever the overall heritability of impulsivity and other personality traits, there are certain individuals in whom habit and environmental exposure play the predominant or deciding role. Heritability is just the average genetic contribution to the trait. Where does the gorilla or the king sit? Wherever he wants: in one case because of genetic endowment (the gorilla) and in the other because of environmental endowment and experience (the king).

Some species are far more impulsive than others and others are more reflective or inhibited. As a species humans are not highly impulsive. This is because most of us, and our ancestors who bequeathed to us their genes, are all our lives walking on behavioral trajectories that require a high degree of impulse control. There are many ways and places for a human to fit into society but all of the landscapes of socially acceptable human behavior are knife-edged. One step too far to the left or the right can result in social failure, ostracism from the group, and worse. Humans are adapted to live in social groups and as a result each of us has a mix of impulse, and impulse control. The overly impulsive individual is likely to be become involved in social conflict and aggressive interactions. As will be discussed, what first looks like aggression is often really impulsivity. Also, the ability of humans to choose multiple pathways sets us up for failure unless we use our ability to make long-range plans. We have to go beyond instinct to plan for our futures and to avoid pathways that may in the short term be continually rewarding, but in the long term disastrous. As we will see, sometimes the problem is that a person has chosen a pathway for which they are not well suited. In the movie *Full Frontal*, one of the characters disparages another: "She drives a car that does not suit her," by which she may have meant that there was a discordance between that person's personality and her life choices.

An obvious example of the need for humans to maintain impulse control over the course of a lifetime, and the consequence of even rare missteps, is addiction. All modern humans are potentially exposed to drugs and other addictive agents, for example gambling and cyberworlds. In the short term these agents produce repeated reward, but in the long run can lead to disastrous consequences.

Impulsive individuals find it considerably more difficult to "just say no" to addictive agents, and once addicted are further impaired in their volitional capacity. As discussed in Chapter 3, one human, or mouse, may be more impulsive or more cautious than another, even owing to the action of a single gene such as the htr2b serotonin receptor.

Considering the effects of such variation in impulsivity in the simpler context of the life of a mouse, this variation in caution and impulsivity has obvious value in different situations that the mouse may encounter. For example, the incautious mouse is more likely to wander into an open field to be borne off in the talons of an owl or the jaws of a fox. On the other hand, the too-cautious mouse may not feed and thrive as vitally as the "moxy mouse" with just the right dose of chutzpah. There should be no doubt that whereas the primal behavioral characteristic – impulsivity – may be the same in man and mouse, the range of consequence is vastly different in the two species, as is the decision making, as we will see next. Impulsive humans, unlike mice, are exposed to a vaster range of risks, and many of the risk exposures are new ones to which there was no possibility of evolving instinctive protective responses.

ZERO-TRIAL LEARNING

Rather than relying on instinct, people choose life paths, but as we proceed along those paths we are constantly, even if partially, aware of the need for self-control, forethought and sublimation of immediate wants. In this regard our internal mental state and the processes leading to the activation or inhibition or delay of action are to some extent and in many situations thoughtful and explicit, rather than unconscious, reflexive and automatic. Placed in an uncomfortable position – like a cat on a "hot tin roof" – we may choose to endure the suffering because of an understanding that the alternative is worse.

It is important to take notice of the fact that much of the behavior of other species is also non-instinctive. Inherent to the adaptive success of most animal species is the ability to learn. This frequently includes the ability to learn by emulating the behavior of another animal of its species, or related species – as in mixed flocks of birds. However, compared to humans, the learning of most other species is far more likely to be associative. These species are hampered by both their more limited conceptual range and their more limited language. Lacking these tools, a mouse may be trained to stay on the experimental apparatus that is the equivalent of the "hot tin roof", which is a hotplate, rather than to jump. However, it takes repeated trials of training to create the associative learning linkage in the animal's brain. It is comparatively difficult for one mouse to simply explain the situation to the next mouse.

Associative learning is not the most important way that humans learn and regulate their own behavior. We often learn by example. Furthermore, we often do not need to see an example or be told. We can build a conceptual model of the situation enabling us to anticipate and avoid the consequences of things that have never been directly experienced. Such zero-trial learning, enabled by the much greater cognitive predictive capacity of the human brain, sets humans apart, or can, from other animals. Humans do not have to destroy the world's ecology via global warming, because we can anticipate the consequences of our actions and modify them.

IMPULSIVITY AND AGGRESSION IN CONTEXT

Impulsivity is strongly tied to aggression for the simple reason that much of a human's environment does not consist of things but is a social network composed of people. If a person is prone to "lose it", soon others will experience the effects. The aggression may be physical but it can also be verbal or emotional. Thus, aggression is socially defined and significant in the context in which it occurs: the living room, the boardroom, the road, the playing field, the hunting ground, the battlefield. In different contexts the same behavior has a completely different meaning, and indeed impulsivity plays a very different role. The soldier who bayonets an enemy in battle is following his training, and may have made himself a hero. The inebriated man who has put a knife into his best friend has just made the biggest mistake of his life.

On an individual level, competitive aggression causes us to excel and accomplish, and for many people is an essential part of existence. Without aggression we would not have survived to discover the tale of our millions of years of evolution. To succeed in human societies some level of aggression is required, and the requirement varies depending on the society and all the particulars of one's situation. The role of aggression and dominance has been vastly overestimated by some who believe they are the modern advocates of social Darwinism. As we shall see in Chapter 15, when discussing the genetics of sex, the meek may inherit the Earth and quite often do. The girl may be more interested in the guy who builds the house than that other fellow who was just trampled by a mastodon. As will be discussed in Chapter 13, behavioral selection is both frequency dependent and context, that is, niche dependent. Regarding frequency dependency, it is often good to be a warrior if all around you are worriers. Regarding niche dependency, it is one thing to be a pacifist living in a gated community, but it is quite another to attempt to forswear violence while eking out a hardscrabble existence in a place where the hand of the law is only occasionally seen.

Selection for aggression works both at the societal level and within societies. The prophet Jesus counseled people to turn aside harshness with soft words. However, it is as well that the non-aggressive and non-violent states and people who live within these nations sometimes have someone else to protect them, because otherwise too many of their voices might have been silenced. Until that happy time when everyone behaves nicely, it is acceptable to pass laws regulating behavior and to have a justice system and police to enforce it, at times using aggressive tactics but never using aggression in disproportion and always searching for ways to defuse rather than inflame. The idea is that force properly applied can reduce the need for force, and sometimes that even works. Thus, when we speak of someone being "aggressive", we are usually making a judgment that their behavior was disproportionate to the social context. What was viewed as "aggressive" in the house I grew up in may have been very different than the experience of another person in another family, society or time. Impulsivity and the reciprocal ability to inhibit aggression or other behaviors, such as how loudly one speaks or whether one touches or closely approaches another person, are attributes that have to be judged in a social context.

By now it may have become obvious that aggression is not a very good phenotype for genetic research, being so strongly influenced by context. There are many types of aggressive behavior and even within a culture the meaning of aggression is strongly dependent on context. For example, rough and tumble play is normal and healthy among children at a playground or adults participating in various sporting events. Rough and tumble play is generally frowned upon in the courtroom or office setting. Inability to participate in such aggressive activity as a child often predicts other problems, and to some extent there is probably a cause-and-effect relationship to that, which leads to every parent's dilemma as to whether they should let their child experience or protect them from the accompanying risks.

MEASURING IMPULSIVITY AND AGGRESSION BY LIFE HISTORY

When the person is old enough to come to the attention of the criminal justice system, assessing whether they have been convicted of violating a law is procedurally much easier than measuring whether they have an aggressive or impulsive temperament. Therefore, criminality is a crude measurement of behavior, but produces data that are more reliably and easily collected than some of the procedural assessments of impulsivity and aggression that we will discuss. Also, criminality represents the expression of predisposition over years of a person's life, albeit an expressed behavior contaminated by many other factors. Some genetic studies on aggression have studied criminal violent behavior, and there is evidence of moderate heritability. In other words, it is true that criminality tends to run in families and that genetic factors are involved. A similar approach to measuring the "genetics of aggression" is to measure lifetime aggressive behavior. The Brown–Goodwin scale is one way of doing that. Similar to recording criminal convictions for aggressive or impulsive behavior, the Brown–Goodwin scale measures *outcome* rather than *propensity*.

In the 1980s, the Brown–Goodwin scale, used in the MAOA by testosterone interaction study in Chapter 4, was created by a colleague and collaborator of mine, LaVonne Brown. He had conducted pioneering studies on the relationship of biological factors, including serotonin, to aggression. LaVonne is a gifted child psychiatrist and conceived the idea that instead of just asking people questions about their propensity for violence he could learn more by measuring their history of it – because impulsive and aggressive people accumulate many such incidents and usually will provide that history if asked (Have you ever hit a person? How many times? etc). "Goodwin" is Frederick K. Goodwin, an outstanding psychiatric researcher and former Director of the National Institute of Mental Health (NIMH). When I was an NIMH Clinical Associate, Fred Goodwin was the leader of the institute, and he was (and remains) a profound influence on the field, having assembled an amazing and probably never-to-be-matched concentration of the world's best psychiatrists and neuroscientists.

MEASURING IMPULSIVITY AND AGGRESSION BY EXPERIMENT

More recently, the assessment of impulsivity and aggression has been greatly advanced by experimental measures that allow direct measurement under controlled circumstances. This is similar to how the impulsivity and novelty seeking of the htr2b gene knockout mice were measured. The mouse, or the person, is placed in some sort of well-defined experimental situation, where as many extraneous variables as possible can be controlled. The brain can be imaged during such testing, revealing predictive differences in regional metabolic activity. These differences seem to be telling us that some people's brains work differently, and localize brain regions that are important in inhibiting impulse. Other brain imaging studies involving probes for response to emotion and reward have revealed that people also differ in the activity of neural systems that mediate emotion, fear and craving. Thus, there is a balance between stimuli and brain systems that create behavioral urges and systems that enable us to moderate, delay, sublimate or suppress these urges.

The Point-Subtraction Aggression Paradigm

How aggressively would you behave towards an anonymous individual placed out of sight in an adjacent room? At the University of Houston some two decades ago, Don Cherek set out to test this behavior in people, and discovered that in an experimental setting many people with aggression and personality disorders would react very aggressively and even initiate aggression.

In the Point-Subtraction Aggression Paradigm (PSAP), each person who takes part is placed alone in a room and given the option of pressing a button in order to accumulate money or pressing a second button to subtract money from a person in an adjacent room. This experiment, like the famous Milgram experiment in which participants were induced by the scientist in the white coat who was actually physically present to deliver shocks of increasing intensity, was in fact deceptive because no one is in the adjacent room. In the Milgram experiment, 50 percent of people could be induced to deliver shocks of lethal intensity. They would do this despite hearing the anguished cries of an actor following the shocks they thought they had administered. That was a disturbing result, with broad implications for people's willingness to follow orders and abdicate responsibility.

The Milgram experiment opened up many eyes to the possibility of ordinary people doing evil, but people's behavior in the PSAP is in its own way more disturbing. Only money is involved, but on the other hand they are following no one's orders and there is no white-coated authority figure at their elbow urging them on. At certain times in the PSAP the computer program subtracts money from the person taking part in the experiment. Having suffered a loss, it is then one logical strategy to stop hitting the money button for a few moments, and engage in a little "tit-for-tat". Most participants will retaliate by subtracting money from the person who was supposedly in the next room.

It is not irrational to retaliate. As proven by game theory, it can be a reasonable way to moderate others' behavior. In real life, and for all of us who are not saints, it is not always best to "turn the other cheek". Measured retaliation is even known as the "tit-for-tat strategy". However, this is not all that happened. Certain people would become highly irritated and retaliate massively, responding disproportionately to the provocation – "tits-for-tat", one might say – and ultimately to the detriment of their own ability to win money. Such behavior is highly reminiscent of revenge and vengeance behaviors that have led to and maintained needless ethnic and political conflicts in most parts of the world. There is something within people that makes them want to strike back hard, or even disproportionately. As Gandhi memorably said, "An eye for an eye and pretty soon all the world is blind." However, these are people who believe that if you take one of ours we should take ten of yours. Even worse, and as Larry Siever, my colleague at Mt. Sinai and a world expert on impulsive and aggressive behavior, observes, some people engaged in the PSAP immediately and with no provocation immediately begin to aggressively attack the person they think is sitting in the next room. Asked why, one responded that he did this to show that he "meant business".

In the world's social calculus, it is an unfortunate fact that we have to put the avengers and the intimidators into the equation. Indeed, if all the rest of us insist on retribution that is exactly equivalent to the offense, the only possible outcome is that all the world will become blind. Despite the human desire for revenge, justice has to be tempered and perhaps the key achievement of civilization has been that punishment and retribution have been placed in the hands of cooler heads who are motivated more by reason than by passion.

Both the PSAP and the Milgram experiment involve deception. The participant is told one thing, and then only briefed afterwards about what really happened. In human research, deception should be avoided whenever possible, even if it appears that the information to be realized is important. As the Chair of a human research committee, that is a principle for which I fought. When a researcher lies to a person or is "economical with the truth", he breaches the obligation to allow the human participant maximal autonomy as to whether to participate. Deception ultimately erodes the bond of trust. Today, some of my collaborators in research on aggression and impulsivity are using the PSAP, because it provides an opportunity to monitor brain responses to aggressive provocation and irritation under carefully controlled conditions. In Chapter 9 we will take up the issue of how human studies are reviewed.

Delay Discounting

To what extent does one discount the value of a larger reward with a longer waiting time versus one when a smaller reward might be more immediately accessed? As seen in our study of the htr2b knockout mouse, this is a powerful measure of impulsivity in that animal model. The more impulsive mouse is unwilling to delay gratification for the same interval of time, even though it has been trained to understand that a longer wait will result in a larger reward.

However, it is important to understand that a person's willingness to delay gratification can be affected by more than genotype. It can also be affected by their experience and their perception of the environment. This could happen in the laboratory, but is especially true in real-life delay discounting situations, where the delays are frequently longer and the reward at stake is larger. If one lives in a chaotic environment where it is possible that the holder of the reward may skip town or conveniently forget their obligation, or where the reward is acutely needed in the short term, it may make more sense to put these factors into the mental calculus, and take the smaller but more immediate reward. However, impulsive individuals, whether their impulsivity has a genetic or an environmental origin, have a pervasive tendency to take the smaller but more immediate reward, even in a laboratory setting where the longest delay is only a matter of minutes.

Go–NoGo

What happens if a person is given a task requiring them to respond on certain trials and withhold from responding on others? Do impulsive people have problems withholding responses? It turns out that the answer is yes. Impulsivity measured in even such a simple context correlates with impulsive personality disorders and predisposition to addictions, which are obviously much more complicated behaviors to evaluate. The Go–NoGo test has many variants, including a new "Emotional Go–NoGo" that is being used to explore aspects of emotionality and emotional differences between people. In its basic form, the Go–NoGo task might include 100 trials, 60 calling for a "Go" button press and 40 calling for a "NoGo", inhibition of response. The scientist can obtain several measures from this simple test, including the reaction time and the number of false alarms in response to NoGo signals. It is these false alarms that indicate disinhibition and impulsivity. Males and people at risk for addictions are, among others, more likely to show the disinhibited false alarm responses.

The Iowa Gambling Task

The Iowa Gambling Task was, not surprisingly, developed at the University of Iowa and has penetrated somewhat into public consciousness via its discussion in Antonio Damasio's book *Descartes' Error*. The task involves four virtual card decks represented on a computer screen.

The object of the game is to win money, and when a card is drawn the result may be to add or subtract. The four decks are not created equal. After a time most people figure that out and begin exclusively sticking to "good decks", but even before they are consciously aware of this, the "bad decks" begin generating higher stress responses, for example in terms of galvanic skin conductance when the computer cursor hovers over a "bad deck". In contrast, people with frontal lobe dysfunction will continue to play the "bad decks" even though they are aware they are losing money by playing those decks, and fail to show the stress responses associated with "bad decks".

INTEGRATING EXPERIMENTAL AND LIFE HISTORY MEASURES OF IMPULSIVITY WITH GENES AND NEURAL CIRCUITS

Genes and neural systems play roles in diverse behaviors. Furthermore, impulsivity and aggression themselves have multiple origins at the level of attention, motivation and emotionality, which are the underlying neurobiological domains. The mediating processes are themselves modified by multiple genes. Therefore, if neuroscientists, who are even more impossible to herd than cats, were somehow all organized to desist from research on impulsivity and aggression or if a "ban" on such research were imposed, advances in the area would continue to proceed at about the same pace, albeit with different labeling.

The neural nexus of integration of much work on impulsivity is the frontal lobe of the brain, and its functions. In 2005 the US Supreme Court ruled that the execution of convicted teenagers was unconstitutional, citing their "underdeveloped sense of responsibility". They were right. The frontal lobe, which mediates that cognitive control, is one of the last regions of the brain to complete its development, and genes and interacting environmental factors affect the rate and character of development of this critical brain region. The relative volatility and impulsivity of adolescents and children will need to be discussed further in the context of childhood psychiatric diagnoses involving dyscontrol, irritability, diminished attention span and impulsivity.

Frequently, the frontal lobe is damaged by closed head injuries such as occur with rapid decelerations in accidents involving moving vehicles and blast injuries that many combat veterans have suffered. The frontal lobe is particularly vulnerable to coup/countercoup injury resulting from rapid acceleration/deceleration of the soft brain against hard bone. A safety principle behind auto seatbelts, airbags and crumple zones, and the helmets worn by cyclists, is to absorb the energy of deceleration and to modify the effects of deceleration. Increasing attention has been turned to the protection of the brains of boxers, football players and other athletes because the concussions and knockouts (KOs) they experience are due to decelerating injury, and autopsy studies have confirmed that the damage to their brains is cumulative and lasting. Frontal lobe damage is also part of the global damage caused by Alzheimer's disease, a genetically influenced neurodegenerative disease that afflicts millions, and Pick's disease, a much rarer neurodegenerative disease that, for reasons unknown, specifically afflicts this region of the brain.

Two famous examples of damage to the frontal lobe are Phineas Gage and James Brady. In 1848, Gage was a 25-year-old railroad construction foreman responsible for tamping the gunpowder, sand and fuse into blasting boreholes. Something went wrong and the powder exploded, catapulting the iron tamping bar through the side of Gage's face, out the back of his left eye, through the top of his head, and then through the air about 80 feet further. Later in his life Gage usually carried around the tamping bar, which was about three and a half feet long, and an inch and a quarter in diameter. The bar, which was smooth and had a javelin-like tapered point, destroyed Gage's frontal lobe, although we do not know the exact extent of the injuries. After the accident, Gage spoke within minutes and lived another dozen years. Gage's poorly documented behavioral changes, a story further distorted by the passage of a century and a half, made him a teaching example of the behavioral consequences of frontal lobe injury.

The most well-known modern example of frontal lobe injury is Presidential Press Secretary James Brady. Brady "took a bullet" for President Ronald Reagan. Brady's suffering and behavioral change also had consequences, helping to launch a movement to restrict access to firearms. In 1981, one bullet fired by John Hinckley, Jr. shattered Brady's left frontal bone, damaged his left frontal cortex and lodged in his right frontal cortex, which ended up being destroyed. Brady was partially paralyzed for the rest of his life, but also had behavioral manifestations of frontal lobe injury including personality changes.

Although neuroscience, and in particular the public perception of neuroscience, is strongly influenced by memorable anecdotes (anecdata) such as the stories of Gage and Brady, it progresses through systematic studies. The information base on effects of injury to the frontal lobe, and regions within it, is deep and wide. In the USA, millions of individuals experience some frontal lobe injury, and thousands have participated in systematic and detailed studies. Such studies use imaging techniques and post-mortem autopsy studies on the brain itself to evaluate the extent of injury, and neuropsychological testing – including many new computer-based methods – to assess premorbid capacities, cognitive and emotional performance deficits, and recovery. New capabilities in functional brain imaging have enabled neuroscientists to correlate frontal cortical inefficiency during performance of tasks with behavioral deficits. Behavioral pharmacology, and to some extent genetics, have revealed roles of particular neurotransmitters in frontal function.

Frontal lobe deficits include reduced attention span (attention deficit), poor working memory (short-term memory) and difficulty in executive cognitive functions, the key to which is the ability to switch between cognitive strategies. This leads directly to problems with planning, reasoning and ability to inhibit emotions. Frontal lobe damage often leads to perseverative behavior and to inappropriate aggression and sexual behavior. It can also lead to *Witzelsucht* – "jokiness" – the trait of constantly, and often inappropriately, injecting humor into situations and the telling of pointless stories. Thinking about the prevalence of some of these behaviors, we can see the danger of neuroscience by anecdote. How easy it is to identify someone who is "uninhibited" sexually, perseverative (for example, completing a book despite indicators of futility), "witty", or being prone to telling pointless stories.

Weak genetic predictors of functional consequences of effects on the frontal lobe are now known in development and in head injury. In head injury, apolipoprotein E (*ApoE*), a gene that predicts part of the vulnerability to Alzheimer's disease, was shown to predict the cognitive impairments that occur in boxers. In professional boxing the head is unprotected; the head is a legitimate target. A KO is achieved by damaging the opponent's brain. Recently, it has come to the attention of neurologists that football players have a very high frequency of head injury. The reason consciousness is disturbed (the boxer was hit and became "woozy" or a wide receiver was KOed by a "nice hit") or lost is because of deceleration injury to innumerable neural networks, and this more frequently occurs without loss of consciousness or concussion syndrome. Later, cells die. Although frontal cognitive performance tends to be worse in head-injured patients, the combination of head injury and the cognitively unfavorable *ApoE* genotype, and also the catechol-O-methyltransferase (*COMT*) genotype, leads to worse outcome.

ANIMAL MODELS OF IMPULSIVITY AND AGGRESSION

With regard to the inevitable integration of genes into the equation, it is also important to understand the importance of studies conducted in animal models, and to understand the limitations of these models. Animal models, especially those using rodents, have enabled invasive and well-controlled studies free of many of the confounding factors that impede human studies. Anyone acquainted with the differences in temperament between various breeds of dogs would understand that behavior is strongly determined by genetic factors, as illustrated by the following three jokes, two involving dogs and one a fox:

> Joke 1. Customer (timidly) to innkeeper: "Does your dog bite?" … "No?" (relieved) … "Ow! I thought you said your dog didn't bite!"
> Innkeeper: "That's not my dog."

> Joke 2. Mailman (angrily): "I thought you said your dog's bark was worse than his bite!"
> Dog owner: "Wait 'til you hear his bark."

> Joke 3. Mr. Fox (puzzled): "Mr. Scorpion, you promised not to sting me if I carried you across the river. Now we will both drown. Why did you do it?"
> Mr. Scorpion: "Because it is my nature. I am a scorpion."

Unlike differences in behavior between humans, which can be environmentally determined, the behavioral differences between species and breeds is obviously primarily of genetic origin. Of course, animal behavior is not always as clear-cut and predictable as that of Mr. Scorpion. Often, the results are muddled. Many studies conducted on aggression in mice and rats are inconsistent and difficult to interpret. It is hard to understand what is going on in a mouse's mind (probably not much, I would wager) when it is confronted with the unusual challenge of a second mouse being dangled by its tail into its cage or when a second free-moving "intruder mouse" is placed in its small plastic box. The mouse is a poor model for primate aggression, and although the aggression scenarios are controlled, they are highly artificial. The main problem is that these sorts of "make the mice fight" experiments yield different results depending on the test. With one test one strain is more aggressive and with a different test another is more aggressive.

Despite the limitations of the animal models, most of what we know about behavioral pharmacology and neurocircuits involved in reward, learning, emotionality and behavioral control are from studies conducted in the mouse and rat, which it should also be said are from an evolutionary perspective no lowlier than the human, just differently adapted. Furthermore, there are now new insights into the origins of impulsive behavior in rats, in particular from a group based at Oxford University and led by Trevor Robbins and Barry Everitt. The impulsivity has more than one origin, involving both an urgency component attributable to a region of the brain known as the striatum and a control component attributable to the frontal cortex. Importantly, the underpinnings of the impulsive behavior can be traced back to the dopamine neurotransmitter system and its receptors.

However, it has been key to show that in humans frontal lobe damage and frontal function are predictors of impulsive behavior and cognitive control, and to directly access these capabilities experimentally, for example with the Iowa Gambling Task and the PSAP, in people who have suffered injuries to this region of the brain. With the methods that are available we can say that many people are impulsive not just because they are not trying hard enough or want to be that way, but because that is how their brain works. As will be discussed in Chapter 13, brain imaging studies are showing us that the brains of many people with frontal lobe impairment associated with a genetic variant in a gene known as *COMT* are working harder, but not keeping up. At the genetic level, the animal models have also pointed directly to certain neurochemical factors that are important in neural pathways that modulate behavior, and that in turn has led to studies of variation in genes such as *MAOA*, *HTR2B* and *COMT*.

Distorted Capacity II
Neuropsychiatric Diseases and the Impaired Will

"Mental illness is nothing to be ashamed of, but stigma and bias shame us all."

President William Jefferson Clinton

One reason that psychiatric disease is so feared and stigmatized is that several psychiatric diseases impair the capacity for choice. Distortion of capacity – craziness if you will – is a disturbing prospect. The alternative view, which has been called anti-psychiatry, is that we can eliminate the stigma by ridding ourselves of the idea. From the perspective of this intellectual hillock, psychiatric diseases represent the reclassification – or reification – of constellations of socially unacceptable behaviors as entities subjectable to medicalized control. Are psychiatric disorders valid disease entities? To what extent do psychiatric diseases predict behavior, and why? In particular, which psychiatric diseases distort or impair volitional control? In the chapter that follows we will see that having defined diseases that alter behavior, it is unsurprising that these diseases, and their alleged vulnerability genes, have become a focus of controversy. Are we engaged in medicine to help patients and their families, or are we – aided by medications, hospitalization and physical methods such as electroshock, practicing some sophisticated form of social control over individuals who exhibit impairment of capacity, or whose behavior we just do not like?

IMPULSIVITY, AGGRESSION AND NEUROPSYCHIATRIC DISEASE

Neuropsychiatric illnesses that are associated with diminished capacity are of two main types: those that increase the impulse for a behavior and those that diminish people's ability to inhibit the urge to follow an impulse. Sometimes, as with addictions and schizophrenia, both the impulse and the ability to suppress impulse simultaneously come into play. Also, either factor may play a role at one time but not another. We will see that one reason to study the neurobiology and genetics of impulsivity in psychiatric disorders is to come to a more unified and accurate understanding of what is happening in the brains of these individuals. The psychiatric diagnoses are telling us that there is a problem, are manifestations of problems, are creating problems, and are identifying groups of people who in common tend to benefit from the same medical interventions – both pharmacological and non-pharmacological. In thinking about the relationships between impulsivity and behavior we can be informed by these disease classifications, but not bounded by them. Indeed, they are mutable and "fuzzier" than we would like, and a major task of biological psychiatry is to use new tools and information to transform and sharpen the process and categorization of psychiatric pathology.

Psychiatric disorders that can affect competence to choose will be only briefly sketched here because the goal is to understand that there is a broad relationship between diagnosis and capacity, with important nuances. Probably the most important nuance is that there is wide variation in the effect of the same diagnosis on decision making, and even the same person may at some point in their life or in some situations be competent to decide, and in other cases be incompetent.

DISEASES WITH DISORDERED IMPULSE

Most neuropsychiatric disorders lead to disordered impulse.

Depression and Anxiety Disorders

Depression and anxiety disorders are a broad group of diagnoses that afflict a third or more of the population. Several are related in terms of genetic transmission and shared genes for genetic risk and by neurobiology, for example at the level of brain circuitry, neurochemistry and neuropharmacology. Along with the addictions, mood and anxiety disorders make up the bulk of psychiatric practice. However, despite the abundance and causal complexity of these disorders, the clinical pharmacology of these disorders, and psychiatric disorders in general, is not very complicated, for example as compared to disorders treated by a doctor of internal medicine. When I was a resident it seemed that psychiatry was a bit similar to dermatology: "If it is wet, dry it. If it is dry, wet it. If that doesn't work, try steroids." Of course that was never really true of dermatology, where one has to be on the lookout for many specific problems, including skin cancers and signs of other serious diseases; nor in psychiatry, where the problem might well be an endocrine disturbance, drug induced or reflective of an unseen cancer or infection. Increasingly, psychiatry looks more than "skin deep" and accesses an increasingly wide range of therapeutic options. Still, it is amazing how often the words "Prozac®" and "talk to the patient" summarize the therapy of mood and anxiety disorders. Psychiatry has many miles further to go.

The mood and anxiety disorders include major depression, dysthymia (milder depression), panic disorder, generalized anxiety and phobic disorders. They are genetically influenced and environmentally programmed. These disorders of mood and anxiety alter both the affect and the thinking of the person at an unconscious level. This is why it does not really help a depressed person to tell them to "cheer up". As Helen Mayberg and Sid Kennedy have shown, there is a neurocircuitry of depression. In groundbreaking studies in which they stimulate a specific region of the frontal cortex with a deep electrode, Helen, Sid and their neurosurgical colleagues have brought into remission the depression of several patients whose severe depressions had been resistant to all other therapy. During the actual surgery and at the appropriate depth and frequency of stimulation from a probe positioned at a specific location in the frontal lobe, the depression of some of these patients lifted, they remained free of depression for extended periods, and depression returned if the battery failed.

Similarly, there is a neurochemistry of depression. Patients whose catecholamine neurotransmitters were depleted by reserpine, which was used as an antihypertensive drug to treat high blood pressure, began to exhibit signs of depression: decreased mood, appetite, interest in sex and energy. Most tellingly, they began thinking depressed thoughts. In other words, while it is true that depressing events and thinking can make us depressed, it is also true that biochemical depression can drive depressed cognition and other symptoms such as lack of energy, appetite and interest in sex. It is a vicious cycle. Also, through the work of Rene Hen at Columbia we know that the therapeutic effectiveness in depression of selective serotonin reuptake inhibitors (SSRIs) such as Prozac® (fluoxetine) depends on the ability of these drugs to release brain-derived neurotrophic factor, a protein that actually stimulates neuroplasticity and neurogenesis (the production of new neurons) in the adult brain. One of the virtues of antidepressive treatments including medications and electroconvulsive therapy, which works in about 95 percent of depressed patients and has saved thousands of lives, is that these treatments also address the cognitive aspects of depression.

Treatment of depression, with all its modalities including drugs, counseling, hospitalization and electroconvulsive therapy, saves the lives of many thousands of individuals who would otherwise commit suicide in any given year. Unfortunately, and as has been much advertised, depressed patients sometimes commit suicide after treatment, and during therapy as they are actually beginning to recover. Suicide is a choice, but it is a choice that can be distorted by depression. More than nine in ten people who commit suicide have depression or addiction. Suicide is the eighth leading cause of death in the USA, with more than 30,000 suicides and at least half a million attempts per year, and it also has frequencies of epidemic proportion worldwide. Surprisingly, changes such as fencing off "magnet bridges" for suicide, for example the Clifton Suspension Bridge over the Avon Gorge near Bristol, England, have a measurable impact on suicide rates. Suicidal people often move in and out of a decision state for suicide, as should probably not be surprising given the gravity of the decision. A few people who have jumped off the Golden Gate Bridge have survived. One reported that as his hand was letting go of the rail the thought that passed through his mind was "Worst decision of my life." About one in six major depressive patients will commit suicide.

Bipolar disorder, affecting approximately 1 in 200 individuals, seems the most mysterious of the mood disorders. As indicated by its name, bipolar disorder leads to alternating cycles of depression and mania, when the person becomes energetic, hypersexual, grandiose and delusional. In the manic phase there is pressure of speech and flight of ideas, but no disorder of thought except for the coloring of thought by happy optimism and whatever delusions may be present. Manic episodes are more likely to occur in early adulthood, and as the person becomes older episodes of major depression predominate.

One prominent psychiatrist, who of course has studied bipolar disorder, well illustrates the effects. "Geraldo" is usually manic, which adds to his brilliance, creativity and energy. On the other hand he is irritable and grandiose, and not a particularly good sport (except when winning!). Like the Russian grandmaster Bogolyubov, Geraldo might leap atop the chess table, hurl his opponent's queen against the wall and scream, "Why must I lose to this idiot?" Except that Geraldo doesn't have the patience for chess. Entering a restaurant, he often goes from table to table shaking everyone's hand, imparting happy or irritable energy. "Geraldo" is quick to recognize that someone is finding the situation amusing and in the next moment you may experience a sudden and very public thunderstorm.

Many esteemed psychiatric geneticists believe that there is a genetic link between bipolar disorder and schizophrenia (a disease which will be discussed later in this chapter). Clearly there is a borderland between bipolar disorder and schizophrenia, and part of this borderland is an entity called schizoaffective disorder. Also, acutely psychotic patients with either disorder respond to the same antipsychotic dopamine receptor blockers. Finally, there are a couple of genes that appear to play some role in both. However, the virtues of this type of diagnostic conflation, or syncretism, are questionable and it would seem to lead to contradictions in both research and clinical practice. By analogy, the BRCA1 gene plays a role in both breast cancer and ovarian cancer, but no one claims these are the same disease even though the genetic connection has been definitely proven. Breast cancer and ovarian cancer have many other distinct causes and consequences, as do bipolar disorder and schizophrenia. There are several clear distinctions between classic bipolar disorder and schizophrenia. One is a disorder of mood and affect and the other a disorder of thought. Clinically, they "feel" completely different. Bipolar disorder frequently responds to lithium salts. This was an amazing and wholly fortuitous discovery which points to a different cellular molecular pathology in at least the bipolar patients who do respond. That molecular pathology is still unknown, but ultimately it may drive the brains of bipolar patients to alternating episodes of high and low mood.

As a geneticist I am particularly moved by patterns of genetic transmission of bipolar disorder and schizophrenia in families, because cross-transmission would point directly to common causality, and the lack of it to the opposite. In general, there is a lack of cross-transmission of the two disorders. Schizophrenia most strongly predicts familial risk of schizophrenia, and bipolar disorder most strongly predicts risk of bipolar disorder. That would argue that many of the undiscovered genes will work relatively independently, even though there will be some elements that contribute risk to both diseases, in the fashion that molecules such as BRCA1 and estrogen create some modicum of overlap between cancers of the breast and ovaries but the risk profiles and cellular biology of these two cancers remain distinct.

The present state of knowledge of bipolar disorder is that for unknown reasons a sizeable proportion of all human populations has a severe disorder of mood. Frequently, the disorder is so severe as to make them a danger to themselves and others. Then they can be hospitalized, if necessary, against their will. However, what about lesser degrees of mania or depression? Perhaps the manic patient is throwing away the family's life savings. Situations such as this present an ethical dilemma, because the capacities for thought and decision are intact; and yet every thought and decision is colored as if seen through glasses that are tinted blue (depression) or rose-colored (mania).

Like mood disorders, anxiety disorders also color our thoughts in ways that can be highly specific. As someone with mild acrophobia, I can reliably report that someone with this phobia is not going to respond nonchalantly as did the man who had jumped off the top of a skyscraper and who was briefly interviewed as he passed the 59th floor: "How are you doing?" "All right so far." In *High Anxiety* (forget *Vertigo*), Mel Brooks is an acrophobic Harvard psychiatrist newly appointed Director of The Psycho-Neurotic Institute for the Very, Very Nervous. He has been booked on the top floor of the vertigo-inducing San Francisco Hyatt. Probably because of the movie, I never had a problem at that hotel. However, checking around the room at an Arizona airport hotel, I noticed there was a sliding glass door, but no balcony or guardrail. This reminded me of a Jack Handy joke, "Hey, instead of a trap door, how about a trap window? You know, one that if you lean on it you fall out. Oh wait a minute, that's how windows already work." Being "very scientific" I tested the door, and it slid smoothly open leaving me teetering at the edge of a multistory drop. The soles of my feet tingled. I backed away. To close the door I had to approach on hands and knees. As illustrated, an interesting feature of acrophobia, and other phobias, is that behavior is not necessarily prevented, but the choices that are made are distorted and colored by predisposition.

Addictions

The genesis of addictions can be related to impulsivity as well as disordered mood. However, in the process of addiction there is a transition to habit formation and further to compulsive behavior. Addictions are common worldwide, affecting more than one in ten individuals. Addictions occur to many different agents, only some of which are drugs, but it is important to understand that many addictive agents are more addictive than others. As compared to other agents that are pleasurable – food, chocolate, or perhaps horseback riding – these psychoactive addictive drugs release dopamine far more powerfully in the main reward region of the brain, the nucleus accumbens, also known as the ventral striatum. Having learned that an agent is rewarding, both stress and cues specifically associated with the addictive agent create an impulse which can be consciously experienced as craving. Whatever their level of impulse control, it is much more difficult for the nicotine addict to resist the temptation to smoke a cigarette than it was for me to for example give up chocolate, which I had to do for health reasons. The chocolate was pleasurable and I still enjoy cues associated with it, but there was never a strong primary pharmacological effect of the food. In smokers who have been studied using brain imaging methods that can measure how much dopamine is released, we know that the nicotine cue will release dopamine in the nucleus accumbens, the brain's reward center, stimulating craving for the "real thing". As a learned response, the chocolate-lover's brain will also show some of the same responses, but these responses are not initiated or pharmacologically maintained in the same powerful way as the addictive drug. Therefore, addictions can represent an example of both disordered impulse and disordered impulse control. Also, these impairments can be circumscribed to a particular addictive agent and they differ for agents of different addictive potential, and depending on the person's experience with the addictive agent the craving may be severe or mild. To some extent the strength of these individual responses can now be measured by brain imaging, both using specific dopamine receptor ligands that enable the measurement of dopamine release in the nucleus accumbens, as Nora Volkow of the National Institute on Drug Abuse (NIDA) has done, and via the study of metabolic activations of the brain with functional magnetic resonance imaging (fMRI). Already, it is being learned that genetic variants can alter these reward activations, as Vijay Ramchandani, Dan Hommer and Markus Heilig, my colleagues at the National Institutes of Health (NIH), have shown. Furthermore, in work to which I contributed, a genetically influenced circuit involving the nucleus accumbens and a regulatory region of the frontal lobe was discovered by Elliot Stein and Eliot Hong, and weakness of this circuit predicts greater craving for nicotine. Some people are born with greater propensity to addictions and craving, and we are beginning to pinpoint the genes that are involved and the brain circuits they affect.

An important aspect of addictive agents is that many, or even most, are not consequence free. Some addictive agents, for example alcohol, lead to impaired cognitive capacity, and thereby to a more general disinhibition of behavior. When people choose to consume alcohol, they have made a choice that can lead to problems that follow from behavioral disinhibition and impaired judgment. Furthermore, when an alcoholic "chooses" to drink, his choice has been strongly colored by his addiction to the drug.

Obsessive Compulsive Disorder

Obsessive compulsive disorder (OCD) is a common, severe, genetically influenced and usually lifelong psychiatric disease. It is often resistant to treatment. Perhaps 1 in 200 people meet the clinical criteria for OCD, and many more of us have some milder obsessions and compulsions. In contrast with the milder forms of perfectionism and compulsion that many of us experience, and that are useful in many situations in life, the obsessions of OCD are disruptive of normal life and happiness. The compulsions vary from one person with OCD to the next. However, whatever the compulsion – cleaning, checking, counting, or a myriad of others – the obsession and the activity required to fulfill it can come to dominate existence. As anyone who has watched Tony Shalloub in the role of the Detective Monk, there is nothing wrong with the impulse control or judgment of most people with OCD. Also, there is not a generalized disorder of impulse. Like a person with an addiction, an OCD patient experiences a very strong urge to perform a specific act, and feels miserable if they cannot.

Tourette Syndrome

This is a relatively common neurological disease leading to disordered motor movements. These can include facial tics, movements of the extremities and vocal tics. The vocal tics often involve copralalia, which is the exclamation of curse words such as "fuck" or "shit". Like behaviors in OCD and addictions, the Tourette's tics can also be voluntarily suppressed, as well as diminished or even eliminated by drugs that block dopamine receptors. By focusing his concentration, the person with Tourette syndrome can eliminate or reduce the tics, but they reemerge as soon as he relaxes. This is a general problem with compulsions, whether in addictions, OCD or Tourette syndrome. With constant monitoring and conscious effort, the behavior can, with a struggle, be suppressed, but ultimately it tends to emerge.

Common Origins of Disorders of Impulse?

It is important to note that Tourette syndrome, OCD and stuttering all have a tendency to co-occur in certain families. Many geneticists feel that this indicates that these seemingly very different behaviors are influenced by some of the same genes. A key to effective treatment of these disorders seems to lie in the inhibition or removal of the impulse. For example, the impulse to use the addictive agent can eventually fade in the addict. SSRIs help some OCD patients and in those who do respond both the obsessions and the compulsive acts subside, so that the person experiences some cognitive relief. Also, Tourette patients successfully treated with dopamine receptor blockers experience a diminished pressure to make the motor movement.

Schizophrenia

Schizophrenia is a psychotic disorder affecting approximately 1 percent of people. Clearly, the hallucinations and delusions of a schizophrenic patient can lead to bad decisions, in a way that is beyond the control of the person with schizophrenia. It is precisely these delusions and hallucinations that are most responsive to treatment with antipsychotic drugs, the original prototype being Thorazine® (chlorpromazine). Even while psychotic, the schizophrenic patient may have adequate (which is to say normal) judgment and control in many areas. When treated with the antipsychotic drug, these other areas are not much affected, and people without schizophrenia do not experience the same sorts of consequence – they only experience side effects such as sedation and motor impairment. This focal effect of antipsychotic drugs on delusions and hallucinations is how one can be sure that anti-psychiatrists such as Thomas Szasz are incorrect in arguing that schizophrenia is the merely the reification of thoughts and behaviors that are socially unacceptable or inconvenient. Also, the ability of certain drugs and sensory deprivation to reproduce delusions and hallucinations, including the same ones experienced during psychotic episodes, strongly argues that there is something happening in the brain of the schizophrenic person that is beyond their control, and for which help is needed. The objective fact is that there are no space aliens inserting thoughts into the brain of the paranoid schizophrenic, and it makes life much more difficult if one goes around thinking that there are.

However, and this is where it gets tricky, schizophrenia is also a disease that affects judgment and the ability to make judgments. In fact, after successful treatment of psychotic symptoms of schizophrenia, one is left with the impression that the core of the disease has been left untreated, and usually there is a significant impaired cognitive component that remains.

Louis Sass, in his brilliant and entertaining book *Madness and Modernism* – which he sent me after a wonderful day touring Pompeii's ruins – proposed that this difference in the residual thought patterns of schizophrenics was not a deficit, but a gain of function, with its own peculiar consequences. Both modernism and the cognitive madness of schizophrenics are reflexive and recursive. It is difficult to encapsulate Louis' conception in one paragraph, and it does some disservice, but reflexivity is the type of thinking involved in saying, "I think that I think that I think that I think that I am." Or to put it more efficiently, "$(I\ think)^n$

that I am", where n is any number of iterations of "I think". Anyone who ever watched the original *Star Trek* will be familiar with the potential weakness created by the trap of infinite cognitive loops. By this stratagem, Captain James T. Kirk was successful in defeating several alien supercomputers. If only it were so simple. I'm somewhat sympathetic to Louis's idea that reflexivity is a strength, because the idea that certain cognitive styles fit certain niches has great appeal, and in the modern world humans have demands for reflexive thinking that our primate ancestors did not face. Reflexive thinking is on the rise culturally, and may have been favored genetically in our human lineage. Because there is a little bit of schizophrenia in my own family, I am even a little more sympathetic, or perhaps biased. It is pleasant to view one's cognitive style as a plus rather than a minus. However, one consequence of reflexivity is difficulty in decision making. A cognitive hallmark of schizophrenia is ambivalence.

Alas, the cognitive deficits of schizophrenia go deeper than reflexivity. There are problems with the gating of stimuli. Most tellingly, careful studies of frontal lobe function, for example by Danny Weinberger and his group, reveal deficits not only in schizophrenics but also in their siblings who are free of the disease but have some of the same genetic loading. As will be discussed more, the frontal lobe of the brain specifically mediates executive cognitive processes that include not only behavioral inhibition but also switching between different cognitive strategies. Often when talking to patients with schizophrenia it is apparent that they have difficulty in switching to an appropriate cognitive strategy and as a result they make perseverative errors. As they say, the definition of insanity is the expectation that one can do something that has failed over and over again and get a different outcome. As will be discussed in Chapter 13 and in the context of a gene, namely catechol-O-methyltransferase (*COMT*), that moderates frontal cognitive function, this ability can be measured experimentally. When this was done on schizophrenic patients and their well siblings by my colleagues Danny Weinberger and Mike Egan, they made a very surprising discovery, which was that even when the schizophrenics and their siblings were performing an easy frontal task that they could complete as well as the next person, their frontal cortex was having to work much harder to do it. About a decade ago, my son Ariel and I traveled with, or as I told one border guard, "against", the Weinbergers – Danny, his wife Leslie and his son Collin in Israel, Egypt and Jordan. We learned a lot about culture and decision making, but especially valuable were Danny's insights on neurocognitive deficits in schizophrenia and the way that altered frontal lobe function might be crucial in the onset of schizophrenia in adolescence, a time of rapid neurodevelopmental change and adaptation.

Delusional Disorder

Delusional disorder is not the condition of holding a political opinion different than one's own. With a frequency of about 1 percent, it certainly is too uncommon for that description to apply. Unlike schizophrenia, delusional disorder is not marked by cognitive deficits. The problem is the delusion itself, which can create the impulse for the person to harm himself or others. Delusional disorder usually afflicts high functioning people and has onset later in life, although there may have been subtle precursors. Typically, a person with delusional disorder believes there is a conspiracy directed against them, so like schizophrenics they have ideas of reference. Paranoid fears torment them and may lead to all kinds of behaviors harmful to themselves, their families and others as they struggle to prevent poisonings, defend themselves and their children, keep the enemy agents and their devices out of their homes and bodies, and so on. Many people with delusional disorder do not come to the attention of physicians. Delusional disorder can express itself in sexual behavior, both in morbid jealousy and in sexual attachment.

One of the first psychiatric cases I saw as a medical student was a patient with delusional disorder. "Mr. Hilltop" was cared for by a wise psychiatrist named Harry Davis, who practiced and taught at the University of Texas Medical Branch in Galveston. Davis was a sophisticated and highly flexible thinker who dealt with situations for which there was no simple formula or "standard of care". For this reason he had some of the "best cases", and he assigned a few of the most fascinating to me. Mr. Hilltop was a self-made millionaire in the oil industry. In his late thirties he began to suspect that the government was out to get him. It started with a tax audit. As the saying goes, just because you are paranoid doesn't mean someone isn't out to get you. Then he began noticing that people around him were "in on it". Being a Texan he had a gun, but he bought more and began sleeping with them. He constructed a pillbox to defend his property. He was distracted at work – it wasn't easy to keep up a façade when many of his so-called associates were actors and spies. He deduced that his own family was "in on it". If not, why did they ignore the obvious truth? For unclear reasons, Mr. Hilltop placed a revolver against his cheek and shot himself; the bullet passed through his cheeks but left him otherwise unscathed. Treated with a dopamine receptor blocker, Mr. Hilltop's delusions subsided, and although they were always, as he put it, "in the back of my mind", he was able to get back into his normal life. Three decades later, as I reflect on his unusual behavior, Mr. Hilltop's persecutory delusions were a necessary ingredient in the harm he did himself, and could easily have led him to harm others. If he had done so, I believe that a judge should have taken his illness into account as a mitigating factor, or modified his sentence so that the primary cause of his behavior should be addressed. However, it would be wrong to say that he had lost all ability to control his behavior. Throughout his ordeal, he actually had a strong ability to control his impulses, and outside the sphere of his delusions was as sane as the next person. Also, the choices he made were constantly determined by his individual character and perspective, of which his paranoia was a part, and in the end he did not, like a Charles Whitman, choose to shoot people from some Texas bell tower, nor did he quite take his own life. Instead, Mr. Hilltop made a completely individual, bizarre, and I think unpredictable, choice to wound himself.

DISORDERS OF IMPULSE CONTROL

The reciprocal to impulse is impulse control. There is probably no other area of psychiatric diagnosis where there has been so much inconsistency, change and confusion as in the disorders of impulse control. Yet, there are millions of children and adults who have these disorders, there are effective, life-changing and even life-saving interventions for individuals who are accurately diagnosed. Also, there is overwhelming evidence that genetic and other neurobiological factors play a strong role in their genesis. Disorders of impulse control start in childhood with attention deficits, hyperactivity and irritability, and begin later in life with neurological syndromes associated with brain injury. We have already discussed the impact of brain injury and neurodegenerative disease. As we learned, damage to the frontal lobe plays an especially important role in releasing behavior, and in leading to impulsive choice. However, the development of the frontal lobe is not complete until early adulthood. This is a powerful reason why behavioral choice is different in children, and why they are treated differently when they are convicted. It is a reason why adult disorders of impulse control are so likely to have their onset in childhood. Finally, it is a reason why the childhood diagnoses of impulse control, hyperirritability and "bipolar", all of which are likely to lead to the long-term treatment of the child with psychoactive medications, are controversial.

Antisocial Personality Disorder

Antisocial personality disorder (ASPD) is the psychiatric diagnosis that has been classically associated with aggression and impulsivity, but its definition has been the most mutable and confusing. A hallmark of ASPD is preexisting childhood conduct disorder – some children with childhood conduct disorder grow out of it, but adults with ASPD get their start by lying, stealing, torturing animals, skipping school, attacking other children and in general behaving as little hellions. The criteria for ASPD can be met based on childhood conduct disorder plus past behaviors including failure to conform to social norms, irresponsibility and deceitfulness. With behaviors such as this, it is unsurprising that more than four-fifths of incarcerated criminals meet criteria for ASPD, and that this common diagnosis is not particularly useful for predicting recidivism.

On the other hand, Robert Hare's conception of ASPD was primarily psychological, not behavioral. Hare's Psychopathy Check List scale attempted to measure the inner state of remorseless, alienation from normal social control and lack of conscience. In that regard it was ahead of its day, pointing to a time when with brain imaging and genetic predictors it might be possible to understand behavioral origins. However, at the time it was put forward Hare's scheme ran against the tradition established by the great behaviorist psychologist John Watson, who had emphasized the value of measurable external manifestations of the brain, as compared to explanations based on unmeasured internal states. This is a very important point. Scientists first and foremost study what they can measure and are attracted to classification schemes based on measurements. We may observe that a defendant has a history of childhood stealing, lying, torturing animals, early drug use and sexual promiscuity. Someone who has examined the child may then report a variety of explanations for the behaviors, ranging from lack of conscience to attention seeking, or cry for help.

From *Westside Story*:

> Dear kindly Sergeant Krupke,
> You gotta understand,
> It's just our bringin' up-ke
> That gets us out of hand.
> Our mothers all are junkies,
> Our fathers all are drunks.
> Golly Moses, natcherly we're punks.

The child may convince Sergeant Krupke, but that does not necessarily make it so. Prior to directly accessing activity of brain circuits or genes that influence them, classifications of antisocial and impulsive behaviors that depend on externally observed behavior have worked better.

The ability to identify genes and identify neurochemical factors that influence ASPD, and the ability to image the activity of the brain, for example while the brain is performing tasks requiring behavioral inhibition and executive cognitive control, are arguably bridging the gap between internal states and external ASPD behavior. In that regard, Adrian Raine, my colleague James Blair at the National Institute of Mental Health (NIMH) and others have shown that many individuals with Hare-type ASPD do not have deficits of frontal function and have not lost the ability to control their impulses. Instead, some of these ASPD individuals appear to be "Hannibal Lecters". These are the conscience-free, remorseless and alienated people that met the Hare definition of ASPD. They may become serial killers. Obviously, the deficit in such individuals is not in their ability to plan or defer immediate action in favor of long-term reward. However, there are three other psychiatric disorders that primarily involve a deficit of impulse control.

Intermittent Explosive Disorder

Intermittent explosive disorder (IED) was discussed in Chapter 3 in the context of the "2B or not 2B" story, in which the 5-hydroxytryptamine (serotonin) receptor 2B (*HTR2B*) stop codon was found to contribute to severe impulsive behavior, and even violent, senseless, murders. IED of whatever cause is common relative to diseases such as schizophrenia and bipolar disorder, and it can co-occur with bipolar disorder and other diagnoses. It is marked by extreme expressions of anger, often to the point of uncontrollable rage, and the behavior is disproportionate to the situation at hand. IED outbursts are brief and are often accompanied by signs of heightened autonomic activation such as sweating, chest tightness, twitching and palpitations. Typically, the person is remorseful afterwards.

Borderline Personality Disorder

Borderline personality disorder (BPD) probably affects between 1 in 50 and as many as 1 in 25 people, although it frequently co-occurs with other disorders. Similarly to ASPD, with which it can co-occur, BPD also is marked by powerful, poorly regulated emotionality, in addition to impulsive behavior. BPD patients are often in severe emotional distress. They form strong, but shifting and unstable emotional bonds. Often the emotional attachment is unreasonable and unreciprocated, leading to tragic disappointments. Our understanding of BPD and the advances in its treatment, limited though they may be, is due to the work of dedicated and brilliant psychiatrists. Don Klein, at Columbia University, was a pioneer in showing that pharmacotherapy was one way of helping the BPD patient gain at least some measure and sense of control. Larry Siever, working with Antonia New at Mt. Sinai Hospital, has used brain imaging to measure the differences in the brains of BPD patients. There are extraordinary obstacles in the way of such studies. BPD patients are not convenient to study and in the end, and despite the moderate heritability of BPD, it will probably be discovered that many of these patients suffer from emotional dysregulation because of early-life trauma, and not because of some innate difference or readily reversible neurochemistry. Here, I am reminded of multiple personality disorder, for which we really have no understanding of mechanism, are sympathetic to the suffering of people who have it and are thoughtful of the likelihood that many people who have it probably had some terrible early-life trauma. Yet, and perhaps unlike multiple personality disorder, BPD is amenable to systematic study. Larry and Antonia indeed appear to be making headway, finding that BPD patients have differences in regional brain metabolic activity correlating with their deficits in cognitive and emotional control. Again, the frontal lobe is implicated, but with BPD the ability of the frontal lobe to modulate emotion may also be coming into play.

Childhood Conduct Disorder

Childhood disorders of impulse control are the most controversial. Childhood conduct disorder (CD) is a bit of a catch-all but is the precursor for ASPD in adults. Children with CD lie, cheat and steal. They are likely to be truant, commit vandalism, and engage in early sexual behavior and drug use. Childhood CD also may frequently include individuals destined to be Hare-type psychopaths because these children are cruel and remorseless to both pets and people. However, a major problem with the CD label is that most children engage in such behaviors, especially when they are very young. As we have discussed, children have not developed executive cognitive control to enable them to regulate their behaviors well. This is what parents are trying to help their children learn. Approximately 5 percent of children for whatever reasons – genetic, family environmental, environmental toxins, socioeconomic – establish a pattern of repeated and continuing CD behavior through late childhood and adolescence. Unfortunately, labeling them as CD has so far yielded little benefit because there is no consensus on how to treat it. Clearly, there is a strong role of culture, because rates of delinquency vary widely between different societies. Those who become violent are overwhelmingly male. Other than the Y-chromosome, the role of genetic factors is not well understood.

Attention Deficit Hyperactivity Disorder

Attention deficit hyperactivity disorder (ADHD) is a neurobiologically based disorder. It is moderately to highly heritable. Further out on the frontiers of childhood disorders are other diagnoses such as childhood bipolar disorder. The behavior of children is notoriously difficult to assess, and is usually entangled in the web of family interactions. Most children with the diagnosis of childhood bipolar disorder would perhaps be helped more by receiving another diagnosis, or none. As for ADHD, I do not believe that this is the case. Many children with ADHD have deficits in brain function, and are performing as well as they can under the circumstances. However, the ADHD diagnosis is currently made by behavioral assessment by the parent or teacher on the basis that the symptoms appearing before the age of seven are persistent and disruptive. The behaviors are in three domains: *inattention*, for example forgetfulness, distractibility and losing things; *hyperactivity*, for example fidgeting, inability to stay seated, restlessness and excessive talking; and *impulsiveness*, for example, not waiting one's turn.

Studies have shown that ADHD diagnoses made in this fashion are reliable, but such a behaviorally based diagnosis might be made in a way that is reliable but non-specific to a frontal lobe deficit. It is also easy to see that such diagnostic criteria have led to overdiagnosis. There are many other causes of attention deficit and hyperactivity in children: child abuse, sensory deficits, sleep disorders, discipline and rearing practices, and family stress and disruption among them. That has very important consequences because the attention deficit aspect of ADHD is treatable, in particular with methylphenidate, a medication that in amphetamine-like fashion augments levels of dopamine in the frontal lobe. When child abuse or family stress or discipline are at issue those problems will not be solved with a drug. In ADHD, frontal lobe deficit makes it very difficult to attend to tasks or play activities. One way of thinking about ADHD is that it is a developmental lag disorder, with children with ADHD being three to five years behind their classmates in executive cognitive control. In fact, the younger children within a grade are more likely to receive a diagnosis of ADHD. However, although the frontal lobe continues to develop in children, the attention deficit can continue into adulthood in perhaps 60 percent of cases, with varying degrees of adaptation. This would imply that there is something different about their brains, and the issue is not just developmental timing. The future of ADHD assessment will hopefully embrace neuropsychology, brain imaging and genetics.

One alley for ADHD assessment that appears to be a blind one is pharmacological response representing therapeutic diagnosis, which is anyway a bit backwards because it is obviously better to avoid the administration of stimulant drugs to children who do not require them. It has been claimed that the effect of stimulants to calm children with ADHD is "paradoxical" and that a favorable response to treatment therefore confirms the diagnosis, but this is not the case. In line with the idea that many children and adults diagnosed with ADHD do not have a specific frontal deficit is the fact that the effect of methylphenidate to improve frontal function is not specific to ADHD. That is one of the problems with stimulant drugs. It is tempting to use them, and the price to be paid may be one only to be forfeited in the long term. Today, some

8 percent of Major League Baseball players receive physician-prescribed stimulant medication for "adult ADHD". This epidemic of adult ADHD in baseball players followed the banning of stimulants in that sport. Illegal amphetamine ("greenies") helped hitters maintain the intense focus of concentration necessary to pick up the rotation of a baseball as it leaves a pitcher's hand, and in milliseconds, process whether it is a 98-mile-per-hour fastball or an 85-mile-per-hour curveball. ADHD children treated with stimulants do show improved cognitive performance; however, as Judy Rapoport at the NIMH showed more than two decades ago, so do many normal or even above-average children. Judy is a world expert on childhood psychiatric disorders, including ADHD and childhood schizophrenia. When she administered amphetamine to children with ADHD they substantially improved their cognitive performance, as expected. Because of the controversial nature of studies on children who are normal controls, the children without ADHD in her study were decidedly above average cognitively, being offspring of well-informed parents who were themselves cognitively above average. Nevertheless, the cognitive performance of the control children, who were already doing well, also improved to a similar extent as the ADHD children. The cognitive response of the ADHD children to stimulants is not "paradoxical", although the behavioral response (they may sit still better) can be viewed that way.

While many people manifest improvements in cognitive function following stimulant drugs, we will see later that whether they do can depend on the stressfulness of the situation and the difficulty of the cognitive task, and at least one gene, *COMT*, plays a role. Gene findings such as this, and the role of genes such as monoamine oxidase A (MAOA) and *HTR2B* in impulse control and genes that modulate stress response and emotionality, will help us revisit the role for genetic influence on both impulse and impulse control, but in a somewhat different light. The effects of genes on these other mechanisms will provide further explanation for the heritability of the disorders that involve impulse control.

Overall, the impulse control disorders, and other psychiatric diseases we have discussed, represent difficult challenges for individuals and families who must live with them and for physicians and all the others – teachers, nurses, social workers and co-workers – who help them cope. In deciding whether people have free will, it is crucial to understand that there are disorders that impinge in specific ways on the choices we make. These disorders are endemic among us and our families. However, I will argue that in the final analysis these disorders are impinging factors, not deciding ones, and are also to be embraced as part of our neurogenetic individuality.

8

Inheritance of Behavior and Genes "For" Behavior
Gene Wars

"Life is a perpetual instruction in cause and effect."

Ralph Waldo Emerson

Progress and perspective in behavioral genetics require the wisdom of many disciplines: psychology, ethics, psychiatry and history. However, sometimes the results of bringing together these disparate viewpoints is an intellectual Tower of Babel. When discussions devolve into ritual combat, the best we can hope is that we are "Dee" rather than "Dum", and try to keep in mind that one accrues scant credit for winning a wrestling match with a midget.

Decades into the genetic revolution, we struggle to master a cross-disciplinary language of genetics and to reach consensus on fundamental elements: heritability of behavior, environmentality of behavior, reaction range, and the significance of individual and population variation. Can we agree on some of the cultural, religious, philosophical and ideological influences on behavior genetics and the uses and misuses of genetics? Whatever our backgrounds, our opinions on psychiatric genetics, including more politically charged areas such as the genetics of antisocial behaviors, should be informed by data. To what extent are psychiatric geneticists "following the data" and to what extent are the motivations of their inquiries, and the conclusions, colored by ideological and social agendas? What are the motives of its critics, and are they relevant?

Psychiatric genetics is sometimes a toxic environment in which some geneticists and psychiatrists have been accused of being cogs in a government plot to tranquilize people after reifying "undesirable" behaviors as psychiatric diagnoses. Adding genetics to the mix inflames the concern. Because both psychiatry and genetics have contributed to legacies of racism and genocide, disparities between ethnicities, social groups and economic strata feed the thesis that modern psychiatric genetics serves the stigmatization and suppression of the socially and economically disadvantaged. Can the science be disentangled from the politics, or do we want it to be? Shouldn't science inform policy and philosophy?

The ideological war on behavioral genetics reached a dubious peak in 1995 at the Wye Conference on the Genetics of Aggression. It was a "happening". Even before demonstrators arrived, Irv Gottesman, a legendary figure in the field, had been called a Nazi. Irv performed important twin studies on aggression and introduced to behavioral genetics the concept of endophenotype (an inherited, disease-associated intermediate phenotype). Irv is also Jewish.

The Wye demonstrators were better organized than the scientists. They brought proceedings to a halt (with my slides in the projector), chanting, "Maryland conference, you can't hide; we know you're pushing genocide." A demonstrator shouted the most apt comment of the meeting. A name-badged "scientist" had become agitated even before the demonstrators arrived. Nothing to write home about, but now he had gone off-kilter, yelling and pushing towards a demonstrator, who had the wit to yell, "Get that scientist's DNA!"

THE DEBATE ON THE HERITABILITY OF BEHAVIOR

The picture's pretty bleak, gentlemen . . . the world's climate is changing, the mammals are taking over, and we all have brains the size of walnuts.

Gary Larson

The main issue dividing geneticists at the Wye Conference from historians and philosophers was the heritability of behavior. The argument was substantially complicated by the fact that the geneticists tended to emphasize the caveats and how much was unknown, in addition to what is known about inheritance and genes. Also, the discussion leapt forward to the threatening consequences instead of proceeding stepwise: behavior → inheritance → genes → genetic predictor → identify consequences → deal with consequences. Having worked our way towards the end of the chain, someone would derail the discussion, sending the whole group back to one of the earlier points: behavior could not be defined, it was not heritable, or if genes were involved they could not be shown to be causal. Controversy over the heritability of behavior continues to the present day, with exaggerated claims on both sides despite clarifying popular expositions such as Michael Rutter's *Genes and Behaviour: Nature–Nurture Interplay Explained*, published in 2006.

There was a fairly standard discussion of the heritability of behavior. Worldwide, identical twins are rather common and the frequency of identical twinning is remarkably consistent: about 1 in 250 births. As an aside, it has occasionally been observed that despite concerns about the dangers of human cloning the existence of so many clones has thus far not resulted in calamities.

Not all of us have an identical twin or know we have one, but if we have an unknown identical twin, their behavioral similarities can sometimes lead us back to them or them to us. For example, the identical twins James Alan Lewis and James Allan Springer famously discovered each other at the age of 39. Eerily, both had first born sons by the same name, married and divorced women named Linda, remarried women named Betty, had pets named Toy, chewed their nails to the nub, worked part time in law enforcement and vacationed at the same beach. Another set of twins, Oskar and Jacob, were separated at birth. One was raised as a Catholic in Germany where he joined the Hitler Youth and the other as a Jew in Trinidad and eventually spent time on a kibbutz. Like the legendary Mallifert twins in the Charles Addams cartoon, Oskar and Jacob found remarkable similarities when they finally met at the airport (rather than at a patent office, where the fictional Mallifert twins were depicted as having simultaneously arrived dressed identically and bearing the same bizarre invention). Oskar and Jakob both read magazines from back to front, stored rubber-bands around their wrists and shared a variety of other idiosyncrasies and mannerisms. The complete story of the Minnesota twins is told in Nancy Segal's definitive book *Born Together – Reared Apart* (2012).

GENOME DETERMINES REACTION RANGE

When people observe dramatic effects of extreme environments on behavior it is natural that they should doubt the preponderance of evidence that diverse behaviors, ranging from personality factors to cognitive ability to psychiatric diagnoses, are moderately (30–40 percent) to highly (60–70 percent) heritable. A useful conception – re-introduced at the Wye conference by Marcus Feldman, is that genes lead to heritability by determining a person's "reaction range", a concept introduced to psychology by Irv Gottesman in 1963. Depending on the environment, the behavior can thus be very different even if the person has the same genotype. The reaction range can indeed be very wide. However, the fact that a range of environmental exposures can be imagined does not mean that they happen. Heritability measures what the role of gene and environment is, rather than what it could be. Although the range of potential reactions can be very wide, reaction range also has to be understood probabilistically. Given a certain exposure, some reactions are much less likely than others. Both factors lead to the observed moderate to strong inheritance of behavior: people do not experience many of the more extreme environmental exposures we could imagine, and for a given exposure people with the same genotype tend to respond in certain ways, even if their potential range of reaction is much wider.

Because the environmental factors to which populations of people are exposed are subject to change, reaction range is the main reason why heritabilities are not fixed attributes of behavior. If the environment is changed such that no one or everyone exhibits the behavior then heritability of the behavior will drop to zero. This is true because many heritable behaviors are contingent on exposure, and conversely some heritable behaviors found in only a fraction of the population can theoretically be induced in nearly all of us via the appropriate environmental manipulation. An additional consequence of reaction range is that if environment is varied dramatically and randomly, the environmental variance will increase and the role of inheritance (heritability) will drop close to zero. This is because heritability is a ratio of variance attributable to genetic factors divided by overall variance. This is shown in Figure 8.1.

By increasing environmental variance the effect of most genes is diluted, lowering heritability, since in the short term increasing the environmental variance does not lead to an increased variability in genotypes.

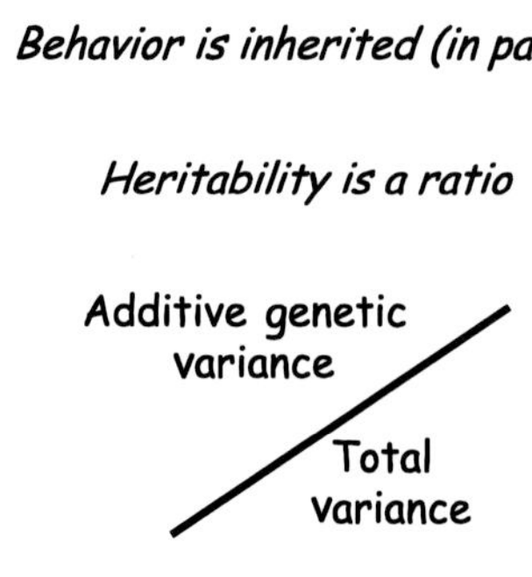

Behavior is inherited (in part!)

Heritability is a ratio

FIGURE 8.1   Heritability as a ratio.

The exception is that some genetic effects only occur following an environmental exposure, so that if the environment is not so varied as to expose some individuals the effect of the gene will not be visible.

REACTION RANGE: THEORY AND REALITY

Reaction range and environmental interaction can also become a clever person's explanation for why a gene *cannot* "cause" a behavior. However, this theoretical argument is trumped by the moderate to high heritability of temperament and psychiatric disease. Heritability does not have magical origins but is directly attributable to genes, most of which are not yet identified. In proportion to their degree of genetic relationship, twins and other blood relatives share a complement of genetic variants inherited in common (identical by descent) from their parents. We do not understand exactly how inherited genetic differences lead to behavioral similarities within the reaction range framework or other theoretical frameworks. However, we should have the humility to admit that this is because we are yet unable to measure many of the specific genetic and environmental factors and also because these factors have not been adequately integrated into the theoretical framework.

TWIN STUDIES: THE CONTROVERSY AND THE METHOD

What do twin studies tell us about the inheritance of behavior, and what don't they tell us? People who oppose the idea that behavior and cognition are substantially inherited do not like twin studies. Many are justifiably troubled by Murray and Herrnstein's book *The Bell Curve*, emphasizing racial differences in cognitive performance (or to put it another way, IQ test scores). These differences were tracked against vaguely defined "race" rather than population. They are more likely attributable to non-genetic factors. Also, the group differences do not tell us very much about the cognitive ability of any individual or the inheritance of cognition. Given equal opportunities, would everyone perform equally well on cognitive tests? There is ample evidence that given equal opportunity everyone does not perform equally on cognitive tests. However, and contrary to the negativistic conclusion of Murray and Hernstein which would restrict human opportunity, many can perform as exceptionally, but can do so under different circumstances and when given individualized opportunities. In a fashion similar to what we have learned of the inheritance of personality and cognition, if we want everyone to do well, some will require more help; and different help, perhaps guided by genotype. Let us explore in more detail the twin method that has generated this conclusion.

The twin method is based on the idea that we can learn about inheritance by comparing the similarity of identical (monozygotic, MZ) twins to the similarity of fraternal (dizygotic, DZ) twins. As mentioned earlier, identical twins are often remarkably similar in behavior, even when raised apart. Fraternal twins tend to be less similar, although resembling each other more closely than two people selected randomly. Heritability can be computed from the ratio of similarity (concordance) of monozygotic to dizygotic twins: the MZ/DZ concordance ratio. For many personality traits and cognitive abilities, this ratio is approximately 2:1 and leads directly to an estimated heritability of 40–60 percent.

However, there are flaws in the twin design. The major flaw is that identical twins experience the same environment. However, the role of shared environment in behavior is not large. The effect of increased sharing of environment by identical twins has been measured in a surprising way. Twins have been studied who thought they were identical but who were actually fraternal twins. The degree of resemblance for personality and cognition of such inauthentic identical twins approximates that of fraternal twins. From the standpoint of the heritability of behavior and cognition, thinking you are an identical twin is not the same as being an identical twin.

If behavior is inherited, then this means that certain genes that twins coinherit from their parents are responsible. This suggests that we should eventually be able to identify how those genes "for" behavior can build a brain and how variations in these genes can influence behavioral differences. Even if the effect of the gene is small and probalilistic, it is the reliability of the gene effects on a population basis that enables evolution to work to act on the frequency of genetic variants by selecting genes "for" behavior. To understand how a functional DNA variant can influence behavior, we will later navigate the route from DNA to molecule to cell to brain to behavior. At that point, we will start with DNA and its natural variations, and the process by which DNA is transcribed into RNA and then decoded by cells to enable the synthesis of proteins and eventually the complex structure of brain circuits that enable behavior.

THE DEBATE ON GENES "FOR" BEHAVIOR

Apart from the problem of heritability, there is the problem of the relationship of the heritable elements (the genes) to the outcome. During the Wye proceedings a philosopher argued that it was always wrong to refer to a gene as a "gene for" behavior (a statement that is theoretically seductive, but false). The argument didn't stop there: supposedly no gene could said to be a "cause" for anything. Getting rid of "cause" was an idea that did not gain traction: I said (and the statement was memorable to me) that if we were not careful we could next find ourselves doing away with "effect", at which point we might as well retreat to caves and convince ourselves that being cold and fireless was just a state of mind. Paraphrasing Leon Trotsky, everyone has a right to be stupid, but not to overuse the privilege. A generation after the Wye Conference, and as I discovered in discussions with a coauthor on a might-have-been textbook, there is still vigorous debate as to whether there are genes "for" behavior. There are two sides to the argument, but the crucial issue is whether evolution has selected certain genetic variations for the express purpose of altering behavior, thus making these "genes for". We will discuss examples of such genes, including the *COMT* "warrior/worrier" gene, later in the book (see Chapter 13). However, in one sense the philosopher and the coauthor were both correct. There are many genetic variations that did not evolve for the purpose of causing a psychiatric disease, but that contribute to disease if combined with an unfavorable mix of other genes and environmental exposures. For example, it is perfectly obvious that there is no gene "for" cocaine addiction, in the sense that humans were only recently exposed to cocaine, and no advantage of being a cocaine addict has been identified.

PEOPLE ARE NOT MONKEYS

A controversy that almost derailed the Wye Conference before it was convened, but that was not directly dealt with at the meeting, was the use of animal genetic models of aggression, and extrapolation (charitably) or conflation (uncharitably) of findings in the animal models to behavior in people. As already discussed in the context of the "2B or not 2B" stop codon (Chapter 3), given the limitations of studies in humans, it is very important that we understand the importance of animal models. They can provide a key to the validation of genetic behavioral findings, including the overall role of inheritance and the roles of specific genes.

I have primarily used animal models to follow up genetic findings made in people. Man-made models such as the htr2b gene knockout mouse can provide an opportunity to test the predictive validity of observations made in people. Also, naturally occurring genetic variants in monkeys and other species such as mice and monkeys can be extremely valuable. At times these sequence variants found in other species are *orthologous*, that is of the same evolutionary origin and function. The rhesus macaque monkey has naturally occurring gene variants at monoamine oxidase A, the mu opioid receptor and the serotonin transporter that are orthologous to or closely mimic the functional variant found in humans, and the list of natural and man-made genetic variants available in other species that are informative for human behavior is constantly growing. In such animal genetic models it has been possible to confirm gene by environment interactions, particularly gene by stress interactions, in a way that would have been impossible in human studies. There is also an overarching and foundational role of animal studies in neuroscience that is sometimes not appreciated, and this is that most of what we know about the relationships of brain neurochemistry and structure to behavior has been learned from studies in animals. It is true that brain imaging and other psychophysiological measures are increasingly enabling access to the human brain; however, this access is, if anything, lagging further behind what is feasible in animal models, because of a variety of powerful new methods that can be used in other species, which frequently depend on the use of genetic tricks. For example, the "connectome" images of Jeffrey Lichtman shown in Chapter 11 and which are enabling neuroscientists to visualize and explore the connections between neurons in an unprecedented way can only have been generated in animal models where the genetic manipulations are feasible and ethical. Similarly, the optogenetics revolution pioneered by Karl Deisseroth, in which specific neurons and circuits are switched on and off using light, depends on the ability to perform a genetic manipulation and to place a probe into the brain. The reasons for the widespread exploitation of animal models in neuroscience therefore do not include either a desire to experiment on animals or overoptimism about the ability of animal models to replicate the human experience.

For many of the higher order behaviors of humans, it is clear that animal models fall short. People are not mice, or monkeys. Emphasizing the difficulty of establishing animal models of behavior, there are no good animal models for several psychiatric diseases, including schizophrenia and bipolar disorder. Mice and monkeys can be induced to drink to inebriation and become hooked on drugs of abuse, and much of what we know about the neurobiology of addiction has been learned from these models. However, the alcohol-related behaviors of a rodent or monkey are not equivalent to alcoholism, which is a human phenomenon that is social in its contexts, expression, diagnosis and course, and human genetics in the sense of the particular functional sequence variants that humans have that influence onset and course. The problem of using other species as models of human behavior is emphasized by the fact that even one person's behavior has a different origin than the next and thus there is no one model for human psychiatric diseases. My schizophrenia or normalcy definitely has a different constellation of causes than does yours, even if we superficially resemble each other. We are not like cars of the same model manufactured in the same plant. Under the hood, our engines are different. Therefore, if an animal model should somehow exactly capture the features of your situation it will probably fall short for mine. While we can learn much about the basic neuroscience from animal models, we constantly return to the human for genotype, context and an understanding of outcome.

Knowing the sophistication of Fred Goodwin, former Director of the National Institute of Mental Health (NIMH), in these issues, he clearly was aware of these nuances and limitations. However, bad things can happen when a scientist speaks in shorthand and in public forums. Fred Goodwin's downfall as Director of the NIMH came unexpectedly, and as so often happens, from a combination of good intentions and getting too far ahead of history. Arguably, hubris played a role. Goodwin had fostered the Violence Initiative, a loosely associated group of studies aimed at understanding the origins of violence, which as discussed elsewhere (Chapter 5) is a behavior that is strongly context dependent in its meaning. In 1992 he was quoted as comparing youths living in the inner city to "hyperaggressive" and "hypersexual" monkeys living in a jungle. Without a doubt his aim was to state that something could be learned about aggression by studying non-human primates – but the result was disastrous. Monkeys are in fact an important animal model of aggression: their behaviors are closer to those of humans than the behavior of rats, with aggression exhibited in some of the same social contexts. Also, the aggression and impulsivity of monkeys is increased by some of the same factors that increase human aggression and impulsivity: low social status, deprivation, maleness and low levels of a particular neurotransmitter, namely serotonin.

Marie Asberg at Sweden's Karolinska Institute, and Markku Linnoila and Matti Virkkunen at the National Institutes of Health and University of Helsinki had found that low levels of a serotonin metabolite, 5-hydroxyindoleacetic acid (5HIAA), in cerebrospinal fluid were associated with higher levels of aggression and impulsivity, and especially as evidenced by suicide attempts. Similar correlations between 5HIAA and impulsive and aggressive behaviors were seen in monkeys. Years later, but too late to play a role in this controversy, it would be shown that gene by environment interactions important in human impulsivity also occur in monkeys exposed to early-life social stress under controlled conditions. Perhaps it would have made a difference to the outcome of the controversy if some benefit of the research, namely the importance of preventing early-life stress exposures, had been shown. These gene by environment interactions, and the measurement of their effects, will be discussed, but the end of the story at the time of Goodwin's resignation was that the field had scales for measuring aggression and impulsivity, a solid biochemical predictor (5HIAA) and models for neurobehavioral research on aggression and impulsivity. However, a blow had been dealt to the credibility and ideological basis of the research.

THE POLITICS OF BEHAVIORAL GENETICS

When demonstrators burst into the Wye Conference center, the leader carried a red flag and yelled, "Workers of the world, unite. Throw off the yoke of your oppressors!" Most of the scientists in the audience were puzzled. Were we the workers or the oppressors? One may well ask what communism had to do with genetics. However, the answer is quite a bit.

Behavioral genetics played a crucial role in a clash of ideas between extremes represented by communism and fascism. Both misused genetics. The Nazis believed in the superman who through innate talent and will forged on to victory and ascendancy, overcoming oppressors. The "oppressor" was devious and vile but in the end conveniently weak. Siegfried easily overcame Mime, once Mime's evil was exposed to the light of day. It was no accident that Nazis turned to social Darwinism and eugenics and it was a foregone conclusion that their genetic, and – it should also be noted – historical and cultural investigations, would reveal innate Aryan superiority. From this false platform it was a short and illogical step to begin tormenting and killing the "inferior", who had no right to whatever they had and who despite their weakness and inferiority were somehow a threat, if only because they were weakening the genetic stock of the nation.

On the other hand, Communists, and their so-called "fellow travelers", socialists, in the extreme but often in the main believe that equality of social outcome is possible and desirable. In this view, inequalities of outcome reflect discrimination, stratification and exploitation. To judge equality of treatment, it is sufficient to measure outcome, be it income or a test score. The solution is to manipulate the input until the output is uniform. Equally rich, equally interesting, equally poor, equally mediocre, equally boring. As depicted by George Orwell in *1984* and by Arthur Koestler in *Darkness at Noon*, an ultimate goal of the collectivist state is the destruction of individuality: in the machinery of the state, people are generic and replaceable parts. The logical next step, implemented by Stalin, was the slaughter of the intelligentsia who, after all, should have been easily reestablished from loyal party stock.

The gremlin of individuality was never exorcised from the socialist machine. There was always the problem of outstanding ability and accomplishment. Worker heroes were honored but in honoring them a seed of doubt was planted. The leaders were themselves not the equal of the "masses" in innate intelligence and capability, and in their hearts never believed they were. By their exceptional natures the leaders of these movements posed a problem for a central tenet of their dogma, which it may be said, had already been run over by their karma. Consciously or unconsciously, innate differences between people had registered in them in the same way that even a schoolchild is aware that some of his classmates are fast, some smart, some slow and some stupid. The inconsistency between innate variation and the goal of building the perfect socialist state festered. How, or why, can one assure (or enforce) equality of outcome if capability and predisposition vary? What is good for the goose may be good for the gander, but not necessarily for the peacock.

In the Soviet Union the concern about individual variation, and the threat of the Darwinist view that something was inherited, led to a peculiar mutilation of genetics known as Lysenkoism. Lysenkoism reprised the Lamarckian idea that experiences of one generation adaptively transmit to following generations, perfecting the adaptation of the species to its environment. In Chapter 17 we will see that the modern science of epigenetics has taught us that gene expression and phenotype can be altered by epigenetic change outside the DNA code, and yet we have also learned that the "slate" of such epigenetic change is largely, and perhaps completely, wiped clean in each generation. Although the Politburo is still out on that question, Lysenkoism remains as wrong today as it was in 1950. Also, Lysenko wasn't engaged in genetic analysis at any molecular level, he was simply promulgating science by ukase.

Lysenkoism taught that if a giraffe repeatedly stretches its neck to grasp leaves, its progeny will have longer necks and tongues. This provided a scientific basis to hope that after a few generations of everyone just trying hard all children could have the "right stuff" to be worker heroes or cosmonauts. The main problem for Lysenkoism was that Darwinism had already happened. A central precept of Darwinism was that inferiority is eliminated not by being rewarded but by being weeded out. Here in America, the strongest predictor of the yen to redistribute wealth is wealth. A laudable impulse compels the most successful, who sometimes are like the "lucky winners" described earlier but who often are talented and striving, to lecture the less wealthy. John Updike wrote of a certain type of wealthy person who was always striving for the success of redistributionist policies and candidates, but who never succeeded. More remarkably, and positively, many decide to engage in philanthropy, but of course while retaining a few hundred million dollars, just in case. However, there seems to be something within almost all people, from the poorest to the wealthiest, that makes them strive to accumulate: money, cars, houses, airplanes, boats, "collectibles" of all sorts, and even other people. To change the impulse to accumulate (the selfish impulse), the social system would have to change, but for better or worse, we are stuck with the human impulses we have, rather than those that might best fit any particular political system.

QUESTIONING THE PHENOTYPE: IS PSYCHIATRIC DIAGNOSIS VALID?

At the Wye Conference a foundational critique of behavioral genetics came from the anti-psychiatrists, who argue that the psychiatric diagnoses are not even real, so why study their inheritance or identify genes that contribute to them? If this is one's position, it might be disconcerting to discover that whatever is measured as a psychiatric diagnosis shows moderate to high heritability. However, this is not really true. There are many heritable traits (for example, height, eye color) that are not diseases. Again, it is somewhat inconsistent that geneticists and neuroscientists have begun to identify highly specific genetic and neurobiological causes of psychiatric diseases, but it is not fatal to the argument. Therefore, it is important to take on this foundational critique, which in the end does help impel improved categorization of behavioral pathology.

Critics following the trail of Peter Breggin and Thomas Szasz criticize both the pharmacotherapy and diagnosis of psychiatric disease. Breggin argues that use of drugs in psychiatry is useless and damaging. As discussed in Chapter 7, psychiatric diseases are frequently misdiagnosed and overdiagnosed. Pharmacotherapy can be ineffective or make the situation worse, especially in the context of misdiagnosis. However, while having some theoretical merit, Breggin's position is unsound in the practical sense. By the failure to recognize depression or the withholding of treatment, thousands of depressed patients would die each year, and many more would needlessly suffer. Similarly, failure to medicate children with attention deficit hyperactivity disorder (ADHD) would set them back in their learning, and while it is true that there is a social context of ADHD it is important to recognize that in many cases the uncontrolled hyperactivity would lead to social exclusion. Szasz's analysis of psychiatry's weaknesses cuts to critical flaws in how diagnoses are conceived, and we would do well to take it to heart even while ultimately rejecting its unhelpful nihilism. In Szasz's view, psychiatric diagnoses are the reification of socially undesirable behavior into categories such that those individuals and their behaviors can be "managed". Thus, homosexuality once appeared as a psychiatric disorder in the Diagnostic and Statistical Manual of the American Psychiatric Association (DSM). When homosexuality became a more socially accepted behavior, it was duly removed. As will be discussed in the next paragraph, this is progress. On the other hand, schizophrenia is a disabling condition that prevents the victim from functioning in most social contexts, regardless of whether we would wish to become accepting of it. Recognition of a disability is an important step in an individual's ability to integrate and become more functional, even if no specific treatment is available, and for the patient with schizophrenia treatment is available. It is one thing to be uncomfortable around people with impaired capacity. That is wrong. However, it is also wrong to fail to recognize their problem, make allowances and offer help. The purpose of diagnosis is not to stigmatize but to enable greater specificity in recognition and intervention.

In the DSM, many or all psychiatric diagnoses have indeed undergone substantial modification, and new disorders such as borderline personality disorder and seasonal affective disorder have been added. This process continues with each new version. However, experts in psychiatric diagnosis (nosology) are painfully aware of the weaknesses and inconsistencies, but think that the continuing evolution of the DSM is a sign of the health of the field, and not a defect. Change is inherent to any discipline of medicine, especially one as immature as psychiatry. A goal of "biological psychiatrists" such as me is to remake psychiatric diagnosis to reflect etiologies, rather than the surface appearance of the same clusters of symptoms.

Already mentioned in Chapter 7 was a major change in criteria for antisocial personality disorder (ASPD). Hare's ASPD of only a few decades ago required that the individual be cold and remorseless. Such people are still with us, although there is now no category that specifically suits them. The modern definition of ASPD is functional, ignoring internal state, and focusing on problems with behavior. In addition, the diagnosis of intermittent explosive disorder (IED) has disappeared. Too bad, because some people have violent outbursts, often self-injurious or suicidal, initiated by minimal provocation and for no gain. However, there was always a problem with the diagnosis because if the person exhibiting the explosive behavior was inebriated, then IED was ruled out. In real life, much explosive behavior occurs while people are inebriated and most inebriated people do not become violent. Today there is no category that covers IED, and ASPD is a diagnosis that can be made reliably, but not necessarily with as much meaning as previously. Progress in psychiatric nosology has been slow, or retrograde, and awaits application of new types of information from brain imaging, neurocognitive testing and genetics. In this important regard, the next version of the DSM (V) is unlikely to advance far beyond DSM IV.

Diagnoses matter, and misdiagnosis and overdiagnosis have negative consequences. Children diagnosed with ADHD frequently do not have it, and may have other origins for their disruptive behaviors that are critical to address. These include abuse and neglect, drug abuse, deficits in vision and hearing, and specific learning problems. In children, there has been a recent epidemic of childhood bipolar disorder, except that most children diagnosed with this disorder probably do not have it. Their parents have read about the new disorder and found a psychiatrist to prescribe medication. The drugs often make things worse, usually do not make things better and almost always making things more complicated. Something is wrong with these children, but treatment of childhood bipolar disorder seldom contributes to the solution.

For many people, primarily women, the popularization of multiple personality disorder resulted in its becoming the expression of internal torment. Let us say that a person experiences an early life trauma such as sexual abuse. Surveys, for example by my colleague Mary Koss, have revealed that between one-third and one-half of women are sexually abused, and perhaps half as many boys. Some are less resilient or experience worse or more prolonged abuse or are abused during a more critical time in development. Acutely, the sexually abused child may have expressed their distress in a variety of ways including depression, anxiety, suicidality and other types of behavioral acting out. I believe that *The Three Faces of Eve*, a popular movie about multiple personality disorder, led some people to adopt this mechanism as the expression of their angst. Perhaps many had better alternatives. The multiple personality disorder itself can lead to further difficulties and complications. If people hear accounts of abductions and nasty sexual experiments by aliens, then expect an epidemic of such cases.

Multiple personality disorder is an extreme manifestation of internal distress, but victims of sexual abuse can manifest their internal turmoil in a variety of other remarkable ways including hysterical blindness, seizures (pseudoseizures) and movement disorders (psychogenic movement disorders). Each of these secondary problems is severe, has its own disabling consequences and can lead to inappropriately targeted medical intervention that only complicates matters without addressing cause. Simple treatments, for example with placebos (as a resident I temporarily cured hysterical blindness with a tuning fork), can play a useful role if the root cause, for example continuing sexual abuse, is prevented. Later in life, and as we found in research in Native American communities, the effect of childhood sexual abuse is to increase the risk of several psychiatric disorders, all of which were already common in women. A woman who was sexually abused as a child is two to times more likely to have depression, alcoholism, other addictions, antisocial personality disorder or posttraumatic stress disorder. She is also more likely to have somatic symptoms such as headache and fibromyalgia, and to have psychogenic movement disorders, pseudoseizures, mysterious weaknesses or hysterical blindness. Sometimes the person becomes nearly unable to swallow, with either no or minimal organic cause. Therefore, childhood trauma, and especially sexual abuse, can be regarded as a rising tide that lifts all psychiatric disease ships. The consequences of these secondary psychiatric disorders can be devastating, leading to the transgenerational and lateral transmission of problems with family networks, for example if the result is alcoholism or another addiction. A girl with pseudoseizures or a psychogenic movement disorder can be incapacitated, with her ability to form good social relationships and perform effectively at school seriously damaged. Psychiatrists, neurologists and other physicians can help; however, very little can be easily resolved either with medication or with talking therapies. Therefore, it is never a good thing for people to establish one of these behavioral patterns, each of which compensates, in some very indirect and tortured way, for a developmental insult. The longer the tree grows in its bonsaied shape, the more irreconcilably distorted it becomes, and it distorts the environment around it.

Psychiatric diagnosis is a tool that is flawed and incomplete, but that nevertheless helps doctors help people. Psychiatric diagnosis is a way-station on the path to something better, but for all its imperfection it is better than the nihilistic alternative. The answer is not to roll back two centuries to a darker era when there was little recognition and no help for problems such as anxiety, depression, addictions, eating disorders and psychoses. No serious person can think that it was actually better when they were made village idiots, packed off to asylums, scourged and burned at the stake, or neglected. Each year in the USA, thousands of lives are saved by the diagnosis and appropriate treatment of depression. More than two million Americans have schizophrenia and the majority are treated. Family life with an untreated schizophrenic is difficult, and the schizophrenic is himself suffering. The addictions, including alcoholism, gambling, and various illicit and prescribed drugs, affect over 20 million Americans. These addicts have a relapsing–remitting disease. It is incorrect to say that they are being diagnosed because others do not approve of their drinking. They are suffering. Many are helped by treatment. Most psychiatric diseases are substantially heritable and the neural basis of several is at least partly understood. In the bootstrap operation to refine and improve psychiatric diagnosis – where gains have been slow and have had to be measured across generations – the views of people such as Szasz and Breggin are too far off the mark to represent a starting point. Yet their critiques have value in representing a challenge to dogma and the insufficiency of what medicine has to offer to mentally ill people. Putting it most simply, they are saying we need less, but what we need is more. Having moved past the denial stage, we can approach the challenge of achieving a more complete understanding of the genetic and environmental forces, and neurobiological processes, that underlie psychiatric disease and thereby better prevent and treat them. The focus on process and etiology will not provide all the answers to problems that represent the end stage and late stage of pathological processes, but already it is yielding dividends.

The Scientific and Historic Bases of Genethics
Who Watches the Geneticists and by What Principles?

Stars in my pocket like grains of sand

Samuel Delaney

If Peter Breggin is not to be trusted to watch over psychiatric genetics, who do we trust? This chapter begins with an evaluation that some geneticists and psychiatrists, for whatever cause – neglect, ignorance, ideology, greed or self-interest – cannot police themselves or the constituencies they

Our Genes, Our Choices
DOI: 10.1016/B978-0-12-396952-1.00009-9

represent. Therefore, on whom can we rely? For ethics, the most durable and generally applicable model we have is a system of human research review and protection that was created in response to the exploitation of captive populations during the 1940s, and whose principles and procedures have been found adaptable to diverse ethical challenges, including the protection of people taking part in genetic and psychiatric studies.

In addition to laying out the historic framework of genetics, and the events that necessitated human protection, the meta-purpose of this chapter is to explore how questions of free will and choice are foundational to any system of ethics. This is done by recursion to specific problems in the process and ethics of genetic research and application of genetic findings. The research genethic context is that of modern clinical research, where we treat all individuals as autonomous and worthy of respect as consenting agents. The broader societal genethic context involves respect for privacy and freedom from genetic discrimination. Why should we treat people with dignity, as if they are autonomous agents whose fates shall not be dictated, if they are automatons who are already enslaved to causation? Are our "ethical" practices only an expedient contrivance to perhaps be disregarded when it becomes inconvenient, and as happened before, or do these rest on deeper bedrock?

The ethical conduct of human research is founded on only a few simple principles, but its structure and procedures continually evolve in response to new capabilities, complexities, dangers and potential benefits inherent to modern biomedical research. Arguably, the few principles that are at the core of human research ethics, like Isaac Asimov's famous "Three Laws of Robotics", are sufficient to guide oversight of human research into the far future. The Belmont Report in 1979 stated three "laws": respect for persons, beneficence and justice. Everything that follows, including informed consent, minimization of risk and maximization of benefit, and equality of treatment and access to research, is built on these cornerstones. Although the ethical conduct of human research is not essential for ethical medicine, these are also intertwined: patients expect the right of consent. We will briefly sketch historical evidence that the conduct of human research cannot depend merely on the individual goodwill and judgment of the scientist. We will see that the need for independent gatekeeping also applies to the conduct of science and its funding and publication, with which genetic and all human research is closely intertwined.

STANDARDS OF SCIENCE AND STANDARDS OF EVIDENCE

The quality and validity of science is the first issue to consider in deciding whether any proposed human research is ethical or any finding is robust enough to be used in society, whether in a clinic or a courtroom or by an employer. To evaluate science, the best system we have been able to evolve is peer review. If the results that will be produced are trivial or invalid then the risk/benefit ratio of the project is unacceptable. This is precisely the point that many critics of psychiatric genetics would make: the science is questionable, so any human research performed to investigate genes and behavior is also to be questioned. However, they would probably be less happy with how this is accomplished because the question of what can be done, what can be published, and how valid and applicable is mainly governed by rigorous peer review.

Peer review is a process whereby on a constant basis scientists struggle to overcome and learn from usually anonymous critique. It works surprisingly well despite glaring examples of scientific fraud and misconduct that leak through the barriers. The main problem is work that is not up to standard but that for whatever reason survives the winnowing. Most scientists bemoan the fact that too much error survives, and their concern is healthy and vital: we have to look at the landscape of science, and ourselves in the mirror, and see at least some defects as they are. This means valuing the gadflies of science, even while shooing them away if they swarm so closely as to block our vision. One reason peer review works is that scientists are much better at detecting flaws in what someone else has done and, given anonymity, are not shy in identifying defects. Anyone who has built a house will know that the carpenter, plumber, electrician, painter and drywaller will with little prompting point out defects in the others' work.

Peer review in action: Ann Elk (Free University of Monty Python) develops a theory of brontosaurus, the name of which was inaptly changed to apatosaurus: "All brontosauruses are thin at one end; much, much thicker in the middle and then thin again at the far end. That is my theory. That is mine and it belongs to me, and I own it, and what it is too." This theory becomes known to the general public, but is never (to my knowledge) reported in a scientific journal. Also, it does not lead to a coherent, fundable project. Sadly, Professor Elk has comparatively little impact on the field of paleontology.

Commitment to the integrity of the scientific process and a feeling that everyone should play by the same rules explains why scientists are peeved when mediocre work or a bad idea survives peer review or is reported through the press. Of course it is precisely those studies that make outlandish and improbable claims that are most likely to be highly publicized. Humorous examples might include the "Theory of the brontosaurus", "Neanderthals invented Football", "Effect of negative reinforcement on cognitive ESP ability", "Oprah's roots: Thirtieth generation granddaughter of the Queen of Sheba", "God gene discovered, and it is a vengeful gene". But, seriously, in my lifetime I have lived through the "discoveries" of cold fusion and life on Mars. These examples illustrate that the process of science is self-correcting but it takes time and effort. Scientific review does not eliminate, but reduces, the level of background noise, be it "cold fusion" or proof of "ESP cognitive ability". This is worth defending, and it is logical that people are passionate in their defense of it. As compared to most types of science, clinical research is far more expensive, usually requiring both institutional support and peer-reviewed funding. An Ann Elk-type clinical study purporting to demonstrate ESP cognitive ability recently survived peer review and was published in a psychology journal. However, defective clinical research studies often do not survive peer review for publication, or only some aspects of the results survive. In a field such as psychiatric genetics, researchers worldwide are engaged in a highly competitive, and ultimately self-correcting, race for knowledge. No criticisms of non-expert gadflies approach the difficulties raised by the insightful probing by editors and reviewers of high-level scientific journals.

Part of the suspicion that science and ideology are inextricably mixed is due to the fact that in past times the distinction between data and synthesis was blurry. It was not uncommon for experiment, analysis and speculation to appear in the form of a book. It is a great advantage to be able to demarcate synthesis from results, and while it has undoubtedly cost us some great books and led to scientific articles being organized in a somewhat rigid way (Introduction, Methods, Results, Discussion and, nowadays, 100 pages of Supplementary Material), we have gained in other ways. In writing about psychiatric genetics I have been able to rely on findings "battle-tested" by peer review but interpretation of these findings is a different matter. One may reasonably question individual findings, but trying to undermine the overall genetic and neurobiological basis of these conditions soon forces one to part company with a massive and multidimensional body of evidence. On the other hand, there are other legitimate perspectives on the meaning of that evidence than the theory of emergent free will that I am elaborating in this book. One of my purposes in writing *Our Genes, Our Choices* was to present a synthesis that goes well beyond what would be appropriate for any scientific journal, in terms of scope, tone and speculative content. This book certainly would not survive the demands of traditional peer review, and yet is built on foundations created by peer-reviewed science.

GENETHICS OF RESEARCH: TRUST, BUT VERIFY

Clinical investigators are constantly reminded and educated about principles and practices essential to the protection of human subjects, and the vast majority of investigators are meticulous and ethical. Nevertheless, we rely on third party review and oversight for compliance. The requirements faced in clinical research, which is also called translational research, are formidable. At the minimum, the researcher seeks the sanction of several committees, including a human research committee and a scientific review committee, both of which I have had the pleasure and pain to chair. These committees review human research protocols that define hypotheses, procedures, analytic design, hazards and precautions.

The informed consent is the most important part of a research protocol because it provides the prospective participant with the information they need to make a meaningful decision as to whether to participate, and to realistically anticipate benefits and hazards. The written consent is a component of a process that also includes discussions with the researcher or other staff. Yet, a lot hinges on it, and therefore it is surprising that we do not know the best length, level of detail or format. It can run to many pages and contain several complex procedures, advanced concepts and esoteric terms. Faced with this complexity, many people stick with their preliminary decision to participate, and that decision is based largely on a general understanding of the study and faith in the researcher's integrity and beneficence. Investigators and those responsible for oversight are well aware of the problem, but there is a constant battle between simplification and the need for complexification such that the consent contains everything needed to make an informed decision. Human research committees include people of different backgrounds, including medical, scientific and from outside the medical/scientific establishment who can understand the protocol and consent in all its aspects, and who endeavor to make the consent complete, understandable and meaningful.

Additional oversight is usual. Use of an investigational drug or device requires Food and Drug Administration (FDA) approval. Use of ionizing radiation triggers special review. The study of children or vulnerable populations, for example individuals who cannot provide informed consent owing to temporary or permanent lack of capacity, leads to other special protections. Scientific review is required on the importance of the science and validity of the design: does the study pose an important question and can it answer the question it poses?

Expedience exerts a constant pressure against ethical oversight. The great majority of clinical investigators, even those who are top in their field, passionately hate this process which occupies so much of their time, and which not only delays science and adds to the exorbitant cost but also stops many studies from ever happening. Clinical investigators are generally exceptional in temperament, scientific expertise, ability to communicate and training in human studies (often from an experienced mentor). Nevertheless, much interesting and potentially beneficial research never leaves the starting gate. Many studies that would benefit humankind and some of the research subjects never happen either because resources are diverted or because of barriers that cannot be overcome. Perhaps it is not surprising that as a psychiatric geneticist I find that studies involving genetics and psychiatry are among the ones that attract disproportionate concern. Psychiatric disorders, particularly addictions, are feared and stigmatized, so there is often more care in reviewing a study of schizophrenia than a study of AIDS. Greater attention is paid to so-called "genetic studies" even though nowadays almost every study is in some sense a genetic study, even if no DNA is collected. However, genetic studies involving DNA deserve very careful scrutiny. In no study involving DNA or the collection of a biological sample containing DNA can we ever provide absolute assurance of confidentiality, because the DNA itself can identify the individual. Modern genetic research is rarely marred by violations of patient confidentiality or interest, an exception being the Jesse Gelsinger case (see Genethics of Gene Therapy, below), which was an interventional study. However, as will be discussed next, the history of science proves that intensive review is a very necessary evil, to prevent greater evils, including the erosion of the conception of humans as individually free, autonomous and worthy of respect – foundations articulated and systematized in the Nuremberg Code and later in the Declaration of Helsinki and Belmont Report.

WHY IT IS A CHORE TO GET A HUMAN STUDY APPROVED: A BRIEF LESSON IN EXPEDIENCE

It is a natural inclination to hold the "good" of scientific advancement and the general public welfare above the welfare of the individual. Also, it is a natural myopia to think that one's own research is justified on the basis of the "general good". Nazi scientists, who constituted a major portion of the German scientific research establishment, conducted a variety of involuntary studies designed to test the limits of human endurance. They justified the involuntary suffering of their subjects with the thoughts that their victims were subhuman and that their horrific experiments were for the greater good of those who were like themselves. However, the only good thing that ever came out of these studies was the Nuremberg Code. At the Nuremberg Trials, the defendants were able to point out examples of highly reprehensible studies carried out in other countries, and frequently on unsuspecting individuals and vulnerable individuals: children, orphans, the poor, the mentally incompetent and those socially marginalized by race. There were a large number of these experiments and if the goals might have been laudable – to prove the origin of scurvy or the role of an infectious agent in disease – their ends did not justify the means.

After the introduction of the Nuremberg Code, at Willowbrook State School in the 1950s and 1960s Saul Krugman investigated hepatitis by injecting newly arrived, severely retarded children with different strains of hepatitis. The study was an advance in research ethics in that the parents "consented", but was the research ethical? As detailed in 1995 by the Advisory Commission on Human Radiation Experiments, appointed by President Clinton, in the dozens of studies conducted in the Cold War era, thousands of unknowing individuals including children were injected with radioactive isotopes or even exposed by deliberate radioactive contamination of the air to study the effects of radioisotopes and their elimination from the human body.

ARE GENETIC STUDIES HARMFUL?

For most genetic studies conducted so far it is difficult to show harm, although one can imagine how harm would occur. In 1972 the National Sickle Cell Anemia Control Act was passed by Congress, withholding funding from states unless testing for this disease was voluntary, and discouraging mandated testing which might stigmatize African–Americans. However, almost all genetic studies involve only the collection of information, and the information is not predictive enough to have much of an impact, should confidentiality be breached. Therefore, the risk is mainly in the domain of information rather than effect, and the genetic markers are not yet so informative compared to other indicators such as measurements of levels of virus in blood, drug screening results and computed tomographic (CT) scans. Simply put, the information that someone has a condition is far more powerful than a genetic marker that only weakly predicts the risk of that condition.

GENES, JOBS AND GROUPS: INDIVIDUAL AND CLASS IMPACT

"Oh what a piece of work is man"

Hamlet, **Act 2, Scene 2, William Shakespeare**

Having discovered a genetic marker for disease or vulnerability, what is to stop man from using it in any way that may be expedient? Only rules, based on a premise that individuals are deserving of respect and privacy. No case of employment discrimination on a genetic basis has ever been brought before a court, although the Equal Employment Opportunity Commission (EEOC) successfully settled one such lawsuit. In that lawsuit, the EEOC maintained that the Burlington Northern Santa Fe

Railroad had secretly tested employees for a rare genetic condition that leads to repetitive stress injuries, including carpal tunnel syndrome, and the Railroad was also screening for common medical conditions such as alcoholism and diabetes. In 1998 a lawsuit on preemployment genetic testing at the Lawrence Berkeley Laboratory was decided in favor of the employees. The Council for Responsible Genetics claims to have documented hundreds of instances of genetic discrimination that did not go to court. Although, as pointed out in a paper published in a genetics journal in 2003, these claims are almost all unverified and might be seen in a different way if all the facts were known, the examples are representative of what people are concerned about and are instructive. For example, a genetic test revealed that a boy had fragile X syndrome, leading an insurance company to drop him from coverage based on a preexisting condition, and in another case it is stated that a social worker lost her job because she mentioned that her mother had Huntington's disease, giving her a 50 percent chance of developing it.

Note the nuances. In the Railroad case, the most powerful information collected by the company was that on diabetes, alcoholism and other medical conditions. The yield on the test for the rare genetic condition was probably zero. In the case of the boy with fragile X syndrome, it is not exactly correct to say that the genetic test was "predictive" of fragile X syndrome. The boy already had the clinical symptoms of this syndrome, which has a genetic basis and for which there is a genetic test. He also might have been more crudely diagnosed with fragile X syndrome by other means. In the case of the social worker at 50 percent risk for Huntington's disease due to family history, no genetic test was performed, and in fact a genetic test might have revealed that she was not at risk. However, she was discriminated against because of her family history. In the same way, she could have suffered discrimination because she had relatives with breast cancer, schizophrenia, obesity or alcoholism: all are genetically influenced even though there is no reliable genetic test for any of them. However, this should not make us feel very secure in the long term, looking towards those more distant horizons when such tests will ultimately be available.

THE GENETIC INFORMATION NONDISCRIMINATION ACT

The Genetic Information Nondiscrimination Act (GINA), sponsored by Representative Louise Slaughter, signed into law by President Bush in 2008, and which took effect only in recent months, is the key development so far in the protection of individuals from genetic discrimination. GINA, rather than the genome, perhaps will be seen as

Francis Collins' greatest legacy as Director of the Human Genome Research Institute (NHGRI). The informative full title of GINA is "An act to prohibit discrimination on the basis of genetic information with respect to health insurance and employment", which includes denying insurance coverage or charging higher premiums or making job hiring, firing, promotion or placement decisions. Some gaps remain. GINA does not cover life, disability or long-term care insurance policies and it does actually limit an employee's ability to sue. Most seriously, employers and insurers cannot discriminate based on the genetic finding, but they can do so based on the genetic disease itself, and as mentioned, most diseases are genetically influenced. In many cases the boundaries between genotype and genetic disease will be hard to define. For example, if a person has a genetic marker for an enzyme deficiency, GINA would prevent an employer from selectively refusing to hire them on that basis. Why would an employer do this? Because healthy individuals are more productive. However, what if they had the elevated cholesterol levels and evidence of early atherosclerosis that correlated with that marker? Or what if they had very early signs of Parkinson's disease that correlated with a different genetic marker? Even though no genetic testing had been done, GINA would prevent the social worker from being fired on the basis of her family history of Huntington's disease, because the law prohibits information on one member of a group from being used to provide information about others. So GINA protects us against our family histories.

However, GINA does not protect us from diseases we already manifest in some way, even molecular. We can still lose our insurance or our job based on any analysis of proteins or metabolites that are directly related to a manifested disease or disorder. This is a very fine, and rather illogical line. If a child is found to have sickle cell anemia by a DNA test showing that he is a hemoglobin S homozygote (both copies of the beta globin gene carry the valine amino acid substitution as determined by DNA, RNA or protein analysis), then GINA will protect him from insurance discrimination. However, if a hematologist draws a drop of blood and determines that the child's red blood cells distort into a sickle shape, making the diagnosis of sickle cell anemia, then the child has a preexisting condition. Also, perhaps the child's race or parentage helped tip the physician off. Would the physician also perform a DNA test? Probably. Why not? It is inexpensive and precise. Does the child benefit from having the diagnosis made? Definitely. It could save him much suffering, and could save his life. Does the child benefit from being labeled, for insurance purposes, as having a preexisting condition? Of course he doesn't.

It is basically crazy that for insurance purposes a diagnosis can be made by clinical history and microscopic examination of a child's blood, but not by a "genetic test". However, the problem here is not GINA, which is a stopgap remedy to reduce discrimination. The long-term solution, which the USA has finally moved towards, is a system of healthcare in which people with preexisting conditions have equal access to healthcare. When that goal is fully realized, the method – "genetic" or "non-genetic" – by which the diagnosis was made will be rendered irrelevant, doctors can better get on with using the most efficient and accurate tools available, and individuals will not, because of the expedient need of a third party, be subject to genetic discrimination.

GENETHICS OF GENE THERAPY

The likelihood that a typical research genetic study will have a sig-nificant impact on a patient's health is low, and largely confined to the sphere of genetic information and discrimination, which we have just discussed. However, there is one domain that stands out as an exception, and this is gene therapy studies. In these interventional studies, the con-sequences can be profound, for good or bad, and the temptation to use individuals can lead to abuses.

In 1999 as part of a gene therapy trial, Jesse Gelsinger was admin-istered an adenovirus that contained the ornithine transcarbamylase (*OTC*) gene. Four days later Gelsinger died of massive liver failure. It was already known that two patients had had serious side effects, and the informed consent did not mention that some monkeys who received the treatment had died. Gelsinger suffered from OTC defi-ciency, an X-linked enzyme deficiency impairing the metabolism of ammonia, which is made when the body digests proteins. However, unlike infants with severe deficiency, Gelsinger was a young man in reasonably good health on medications and dietary protein restric-tion. This was because some of his cells were normal, his body being a genetic "mosaic" (here is one of the exceptions; not every cell in one's body necessarily has the same DNA). His participation in the study was therefore altruistic. He did not stand to gain as much from treatment, and furthermore he had high ammonia levels that according to the protocol should have excluded him. According to Gelsinger's father, in Senate testimony, investigators had said "the worst that could happen in the trial would be that he would have flu-like symptoms for a week". The lead investigator and the University of Pennsylvania both had financial stakes that were not disclosed to the research participants and probably not to the institutional review board (IRB), under rules in place at that time. The university later paid the family an undisclosed amount.

GROUP CONSENT OR INDIVIDUAL CONSENT?

In studies I have done on alcoholism and related psychiatric disorders in Native Americans, my lab has always carried out the research in the context of the tribal community, and with group consent as well as the consent of individuals. We made this decision because of the strong group identification of many of the participants and non-participants who might also be affected by the research. This is a dramatically different way of conducting a human research study, is not strictly necessary and is certainly not expedient. For example, many Native Americans, including many who we studied, do not live on reservations. They could be studied under auspices of protocols that do not necessarily target Native American tribes, but which include and identify some of their participants as Native Americans or even as descendants of a particular tribe.

The concept of group consent is potentially applicable to other groups, but it is debatable whether it should be. In similar fashion, researchers study Ashkenazi Jews, Japanese Americans, Chinese Americans and individuals of particular ancestries and cultures. Even if they have not revealed their ancestry, it can be determined via ancestry informative markers, and indeed this is important for the validity of most genetic studies and for many studies that are not on the surface genetic.

A second type of "class" focus that human studies can have, and often do, is selection of participants with a certain disease. Thus, all people with Tourette syndrome (a movement disorder in which there are also vocal tics) can be considered a class, and they might be affected by any discovery relevant to their disease or if someone should write about their disease in such a way that they could be stigmatized. For example, David Comings, an expert in Tourette syndrome and diseases involving behavioral dyscontrol, reported that the gene for Tourette Syndrome makes people hypersexual. The gene and the assertion about sexuality of Tourette patients remain unconfirmed. A third kind of class focus is the family. Whenever someone in a family is studied for a genetically influenced trait, and especially if the study is a genetic study, other members of the family are affected. This is also true if the trait is culturally transmitted. In most family studies a family history is obtained. The person interviewed provides information, including sensitive information on other family members. Those relatives are thus studied without consent.

Finally, there is the dilemma that studies done on any individual affect all of us. This is true because in the genomic era any study performed will be informative for some of what is going on in each of us. This is seen most profoundly in genetic studies, although it is also true in other studies. Each of us carries about three million genetic variations in our three billion DNA nucleotides, and almost all of these genetic variants are found in other people. By studying others we also learn about ourselves and about others in the wider "human family". Often, but unpredictably, we will be specially affected by the new information. It is this dilemma that may eventually put an end to the genetic aspect of the discussion of group consent. The *BRCA1* mutation is far more common in Ashkenazi Jews but it is also found elsewhere. No one group "owns" *BRCA1*.

This last discussion of the weakening of the specificity of impacts of genetic research on group is, however, futuristic. We are not there yet, and even when we get there the first principles of human research and the application of the knowledge are to acknowledge autonomy, preserve dignity and exercise beneficence. By enabling people to give informed consent, they decide issues for themselves even if in our opinion their decision is not the best one. By respecting privacy and autonomy we forgo some uses of genetic information that would represent efficiencies. When it is possible to obtain consent of the affected group, this would seem to enhance that. In a few instances that has happened, but multiple issues are raised by a decision to seek group consent. What should group consent consist of? Should it require a majority of the group? Who should be empowered to give group consent? How does one balance an individual's autonomy of choice with the need to protect groups? In this regard I believe that Native American tribes offer sound alternatives. To represent the "will of the people", the better organized tribes have tribal health committees, human research committees (IRBs) and lawyers. It is an imperfect system but one that is capable of improving itself. In our studies on alcoholism and schizophrenia in American Indian Tribes we augmented this with the use of community focus groups in combination with Tribal Health Committees and the rest, working with tribes on a community basis for up to a year to study and improve the content and methods of studies, including tribal co-investigators.

AUTONOMY: PRETENSE AND REALISM

In human research and in the application of science in the context of a democratic society we treat people and groups "as if" they are autonomous and deserving of respect. Is this a convenient pretense, a temporary expedient until conditions change or a way of buying such treatment for ourselves? Alternatively, are people free, requiring us to treat them as if they are? There are many countries in which freedom is not acknowledged. Punishment for prohibited behavior is not acknowledgment of freedom of behavior; it is behavior modification, social statement and revenge. The reality of free will and individual autonomy will be the major focus of the latter half of this book.

Was it frustrating when we had to start over with a new tribal administration or if council was not in session because bearskins had been taken and a ceremony was to take place? Was it frustrating in the first place to have had to vet the study through the peer review of a human research committee? Of course, but as they say, the bears come with the territory and we are better off having survived the process. However, what if we felt differently?

Is there any fundamental reason to accept the autonomy of others, other than the Kantian imperative to do as one would want done generally, or the situational ethical imperative that Derek Parfit has endeavored to expand into a general imperative, or is it a matter of hearing and obeying "Do unto others as you would have them do unto you." Each of these is an essentially faith-based argument, and if we retrace the circuit of Parfit's argument to its convergence with situational ethics, each is also in the end an inconstant and ad hoc basis on which to construct a moral system for how people should behave towards one another. By and large we are doing the right thing, but we need to do it for the right reasons. We should treat people as free and autonomous, not because we are following diktat or are persuaded to so by the Nuremberg Code, Declaration of Helsinki, Belmont Report, Bible or the type of practical ethics argued by Parfit, but because we are persuaded to do so because of what people are.

If we do not believe that persons are autonomous, our commitment to their autonomy is as unreliable as it is expedient. The moment that the right dictator, war, social crisis or important scientific question arrives and it becomes inconvenient, our "respect" for the autonomy of persons or that group of persons will inevitably erode, because in the first place our moral framework was constructed on a foundation of words. Also, it is illogical to believe that thinking persons who are treating others "as if" they are free will not constantly be behaving "as if" rather "as is". That is exactly what I would do. I would behave "as if" you are free, but as will be seen, we all indeed are. Millions of words have been written and more will be about individual autonomy and free will and the foundations of moral behavior, but this book's approach to the problem will take a science-based path by establishing the neurogenetic roots of individuals' ability to make free choices. Will that be the end of the discussion? Hardly, and hopefully it will excite at least these following attacks: (1) it has all been said before (*and it probably has in one way or another*); (2) he is ignorant of moral philosophy and that which he is criticizing (*not quite true*); (3) many questions in moral ethics are not automatically resolved by this observation (*true*); and (4) people are not free. All behavior is understood as the product of causally constrained automata (*and therein lies the heart of the debate*).

But first we have to build a brain from the instructions in a molecule.

10

The World is Double Helical
DNA, RNA and Proteins,
in a Few Easy Pieces

In the molecules composing our cells is ample genetic variation to distinguish any of us from others, except if we have an identical twin. Most molecules in this universe consist of only a few atoms. For example, a water molecule (H_2O) consists of only three: two hydrogen atoms and one oxygen atom. These small molecules are inadequate for the task of carrying the information that makes us individual. However, surrounded by the water of human cells are some huge biomolecules: DNA, RNA and proteins. Each is a large heteropolymer consisting of hundreds, or even billions, of smaller molecules that have been enzymatically linked together into a chain and each has many forms. In this regard the biopolymers are more akin to synthetic nylon or rayon polymers or natural silk molecules in the fabric of clothing or in the strands of rope than they are to small molecules such as water. Because the

Our Genes, Our Choices
DOI: 10.1016/B978-0-12-396952-1.00010-5

small molecular units are polymerized (chained together) in a particular sequence to form DNA, RNA and protein, all three of these biopolymers also differ from nylon and rayon polymers in a critical way. They represent a highly ordered type of information storage, and that information can be decoded by proteins that form molecular machines in our cells, or by a someone armed with a DNA sequencer and knowledge of the code. Also, although much of the emphasis of genetics is on the decoding of the linear sequence of the DNA, RNA and protein biopolymers, which is their primary structure, there is also much more information inherent in their higher order structures and modifications. This is due to the ways, often variable from person to person, cell to cell or even molecule to molecule, in which they can be altered. They can be secondarily modified, for example by the attachment of methyl groups to DNA or sugar groups to proteins. They are folded in complex ways, which is their secondary structure. There are also other higher order interactions between molecules leading to very large, complex assemblies of different biomolecules. There are tertiary and quaternary structures, and beyond.

To understand how this information in DNA and proteins is read and interpreted we need to "follow the molecules" and can do this in a way that is in part historical. This is because scientists first studied the external manifestations of genes, then the proteins that determined those effects, and finally, after Oswald Avery, in 1944, discovered that DNA was the carrier of genetic information, the genes that encode the proteins. With the sequencing of the human genome, the challenge in human genetics has paradoxically reversed from the level of DNA sequence back to the more complex levels of gene expression and the external manifestations (such as human behavior) determined by gene expression. In the next chapter of this book (Chapter 11), and to close the loop, we will discuss the interactions of these biomolecules with each other, and with the environment, to create the complex phenotypes that are externally visible and that were the original starting point for human genetics.

RECIPES FOR DNA ANALYSIS

To understand how DNA is analyzed, it is important to realize that it is a complicated molecule, but in the end a chemical. This means that if a set of procedures is carefully applied ordinarily there will be a dependable outcome. In other words, and contrary to what some frustrated scientists in my lab tell me, this science is often akin to following a cookbook. For example, here is one kit, using a molecular geneticist's recipe for good Italian food. Step 1: Take one pack of Stouffer's frozen lasagna. Step 2: Open it. Perhaps the simplicity of molecular biology is misleadingly

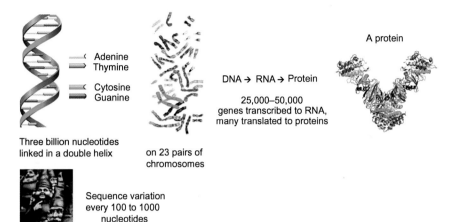

Adenine
Thymine
Cytosine
Guanine

Three billion nucleotides
linked in a double helix

on 23 pairs of
chromosomes

A protein

DNA → RNA → Protein

25,000–50,000
genes transcribed to RNA,
many translated to proteins

Sequence variation
every 100 to 1000
nucleotides
"Gene gnomes"

FIGURE 10.1   Introducing the genome.

overdramatized by this example, but the point is that there is no voo-
doo involved, if you simply do what you do so well. If you add the right
ingredients and follow all procedures, the soufflé will rise.

As shown in Figure 10.1, a human genome consists of some three billion base or *nucleotide* pairs. The DNA chains of humans are comprised
largely of four nucleotides: the *purines adenosine* (A) and *guanine* (G) and
the smaller *pyrimidines thymine* (T) and *cytosine* (C), and these nucleotides
are linked together into long polymeric strands. More recently it has also
become recognized that many of the cytosines have become methylated
and hydroxymethylated, leading to important differences in gene regulation, but although it can be stated that there are at least six relatively
common DNA bases, for our purposes this is a nuance: at this level of
discussion we can say that there are four DNA bases. Human DNA does
not exist as a single polynucleotide strand but is organized into 23 *chromosome* pairs, one member of each pair having been inherited from the
mother and the other from the father. Each chromosome is a long DNA
duplex strand wound around protein cores composed of histones and
supercoiled, compacting it and forming a complex structure critical to
gene expression and to DNA stability, replication and repair. During the
formation of egg and sperm, new combinations of genetic variations that
are already present in the DNA are generated by crossing over between
maternally and paternally derived chromosomes in chromosome pairs,
most human chromosomes being found in pairs. Also, new mutations
constantly occur, although the vast majority of unplanned DNA changes
are corrected by repair processes. In eggs and sperm, one copy of each

chromosome is present, and in other cells there are usually two. Thus, in humans, the *haploid* chromosome number (found in eggs and sperm) is 23 and the *diploid* number (found in the rest of our cells) is 46.

As seen in Figure 10.1, DNA is a double-stranded, twisted ladder with two types of nucleotide pairings between opposite strands: A–T and G–C. The moderately sticky hydrogen bonding between the nucleotides on paired DNA strands allows the DNA to be melted apart into single strands and then reannealed into double-stranded DNA. The melting can be accomplished by raising the temperature, by altering concentrations of ions that are in the water surrounding the DNA or by proteins that can facilitate, as well as inhibit, the opening of the DNA. The ability of the double-stranded helix to split into single strands enables the DNA sequence to be faithfully replicated from a single-stranded DNA template into either RNA, a messenger molecule within the cell, or a new DNA strand, so that the cell can divide and transmit a DNA duplex to each daughter. Note that each daughter cell receives a DNA duplex that is half old and half new. This important consequence of the double-stranded nature of DNA was immediately pointed out by James Watson and Francis Crick when they discovered the structure of DNA, an event that happened so very recently, in 1953. Their insight was to realize that the double-helical nature of DNA would allow it to unzip and replicate copies of itself. As an aside, be careful around such genius. I have heard Watson exclaim (while someone else was speaking), "I don't believe this!" It would also turn out that the double-stranded structure of DNA that Watson and Crick discovered was critical to the ability to probe specific regions of DNA for genetic variation and to artificially amplify and thereby purify selected DNA regions, thus vastly expanding access to human genetic variation.

The study of DNA leads to *functional genomics*. Our DNA contains most of the instructions for building and operating a human, and is contained in the cell nucleus. The nucleus also contains much of the molecular apparatus for the maintenance, repair and replication of DNA and its transcription (copying) into RNA, a single-stranded polynucleotide consisting of *ribonucleotides* (R) instead of *deoxyribonucleotides* (D) because a different sugar molecule is attached to the nucleotide, and substituting *uracil* for thymine. An increasingly larger class of these RNA molecules has been recognized as independently functional, regulating DNA and each other and even serving structural and enzymatic (catalytic) functions. However, much of the RNA is messenger RNA, which will be in turn processed and translocated from the nucleus to the cytoplasm of the cell.

Once in the cytoplasm, the RNA can be *translated* into protein by a large, multisubunit structure known as the ribosome. For cells and tissues of the body, the proteins made in the ribosome and further processed in other cellular structures such as the Golgi apparatus serve as building blocks, signaling molecules and machinery. As mentioned, the same DNA is found in all cells of the body, except most human erythrocytes and platelets, two constituents of blood. Otherwise, hair follicles, white blood cells from blood, brain, fat and muscle all contain the same DNA. The RNA and protein compositions of cells are, however, very different. Cells have different mixtures and quantities of RNA and protein types, and it is the differences in these molecules that build and operate a cell that make a muscle cell different from a neuron.

WHAT IS POLYMORPHISM?

Both DNA and the RNA and protein molecules it encodes are subject to genetic variation, and any of these genetic variations can be measured by sequencing the DNA or by genotyping (a limited form of sequencing) specific sites. The human genome contains at least 22 million common genetic variations (more are being discovered), which are called *polymorphisms*. A polymorphism is a genetic variation with a frequency of greater than 1 percent. However, there are also very large numbers of rarer genetic variants. For 22 of our 23 chromosomes, almost all of us have two copies, thus giving us two chances to have the polymorphic variant (*allele*). The exceptions are the X-chromosome, of which most males only have one copy (inherited from the mother), and the Y-chromosome, which males inherit from their father. Also, many of us have large extra and missing pieces of some chromosomes, an example being Down's syndrome, which is caused by inheritance of an extra copy of chromosome 21 and sometimes by an extra piece of chromosome 21. Speaking of extra or missing pieces of chromosomes, it has recently been recognized that all humans also have many large DNA deletions and insertions, called *copy-number variants* (CNVs). Most of these are too small to be seen using a light microscope but can involve the deletion or insertion of thousands of DNA nucleotides. Such CNVs can delete copies of genes that are encoded by that DNA, or lead to gene duplications as in the chromosomal region implicated in the type of Down's syndrome associated with an extra, translocated, piece of chromosome 21.

For the DNA sequences present in two copies, and except in instances of inbreeding between close biological relatives, there are many alleles that distinguish the copy of the DNA we have inherited from the father from that inherited from the mother. On average and across the three billion DNA nucleotides that constitute our genomes, there is an allelic difference between the maternal and paternal DNA copy once every 1000 DNA bases. Each position that may be genetically variable is said to be a *locus*. If two identical copies are present the locus is *homozygous* but if the maternally derived copy differs from the copy transmitted from the

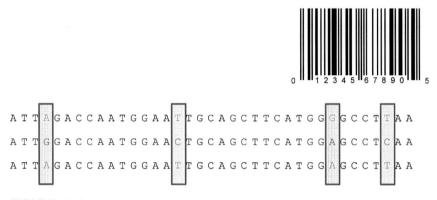

FIGURE 10.2   DNA barcodes: DNA variants in a local region form combinations (haplotypes).

father the locus is *heterozygous*, meaning that the sperm and egg (the *zygotes*) that formed the human were different (heterogeneous) at that locus. Thus, there is not only genetic diversity between genomes, but also diversity within our genomes.

Most of the sequence variants in the genome are of the *single nucleotide polymorphism* (SNP) type. Figure 10.2 shows how different copies of the same chromosome can have different SNP alleles at the same site. Furthermore, it shows a new level of DNA characterization that is increasingly recognized as powerful. This is the combination of several of these SNP alleles in a local region into a *haplotype*, which as the figure suggests is like a DNA barcode that can be used to identify the contents of a DNA sequence in the same way that barcodes in supermarkets can be scanned to reveal what is in a package. For most small regions (packages) of the genome, say 5000 DNA bases, there are relatively few (two to five) common haplotypes, and relatively few SNPs, called tag-SNPs, need to be genotyped to capture these haplotypes, even though the haplotype may consist of many markers. This is shown below. The haplotypes, tagged by a few SNPs, can be used as a type of "supermarker".

Amazingly, the type of genetic variation shown above is not present in some animals, and this level of variation is not found in some species. Animals that have been inbred by the deliberate mating of closely related individuals for many generations (for example, inbred strains of mice produced by 20 generations of brother–sister mating) have lost almost all their genetic diversity. At each locus these inbred animals are *homozygous*, and furthermore all "individuals" from the same inbred strain are genetically identical, like twins. Also, mammalian species on a natural basis differ widely in their overall levels of genetic diversity. Species with very large breeding populations tend to be far more variable. For example, prodigious numbers of red-backed salamander are found in the Maryland woods, and this species is highly variable. Species with large population sizes tend to be more genetically diverse, whereas species with smaller populations or that have suffered population bottlenecks have less diversity.

In general, genetic variation is good: it makes species more adaptable, and in that regard humans are much better off than certain species, such as the endangered cheetah and Florida panther whose dangerously low genetic diversity was discovered by Steve O'Brien. I once helped with a paternity problem Steve was having. It involved giant pandas (*Ailuropoda melanoleuca*) at the Washington National Zoo. Chia-Chia was the mother. However, was Ling-Ling, the London male whose sperm was used to artificially inseminate her, the father? Or was it Hsing-Hsing, Washington's not so studly panda whose ability to successfully mate had been impugned? By studying the pandas' proteins we found several that were useful as genetic markers and heterozygous in the baby panda; this was not easy because the genetic diversity of the pandas' proteins was indeed greatly reduced. Two of these markers could only have been transmitted from Hsing-Hsing, the Washington male. One panda's reputation was saved.

Unlike humans, pandas, cheetahs and Florida panthers have undergone severe *population bottlenecks* such that these species have lost almost all of their diversity and are particularly vulnerable to infectious diseases that at any time could kill the last of their kind. However, because humans haven't always been so numerous, we are much less diverse than some of our ape relatives. If this is not quite clear, then as Carl Merril, a mentor of mine, liked to say, "Not to worry". This was only the first of the three essential phrases I learned under his tutelage, the other two being "Sounds good to me" and "What do I know?" The take-home message or moral of the story of human sequence variation is that our genomes are a trove of sequence variation, and all the cells in our body, save some in the ovaries and testes, have a nearly identical pattern of variation that can distinguish us from all other individuals, unless we have an identical twin.

PROTEIN POLYMORPHISM

As just alluded to in the story of paternity in the giant panda, the first molecule whose genetic variation was studied by geneticists was not DNA but protein. Proteins are encoded by the DNA, but a particular protein is found only in some cells and tissues of the body, whereas the DNA is distributed nearly universally. Proteins are encoded by regions of DNA called genes. In particular cells, these regions are active owing to differences in regulatory proteins bound to DNA, differences in chromatin (DNA-protein) structure and differences in DNA modifications including DNA methylation. Active gene DNA in the nucleus is copied or *transcribed* into RNA, the single-stranded messenger molecule. The RNA is processed and then translated into a protein sequence in the cytoplasm. Protein synthesis takes place at the ribosome, where the message on the RNA is decoded by small molecules called transfer RNAs. The protein is synthesized one amino acid at a time, building a large heteropolymeric (multiple different subunit) protein molecule whose sequence fairly faithfully reflects that of the original DNA molecule. The whole process can be likened to the conversion of sheet music to the music we hear from a player piano. The sheet music (DNA) is transcribed to player piano tape (RNA) and the tape is translated by the player piano into individual notes (amino acids) linked together as music (protein). Marshall Nirenberg, a Lab Chief at the National Institutes of Health who recently died, won the Nobel Prize for being one of several scientists to first decode one of these protein notes, namely the DNA sequence AAA, which when transcribed to RNA is recognized as the triplet nucleotide code for the amino acid phenylalanine.

If one of these amino acids is changed because of a nucleotide change there can be a detectable alteration in the structure of the protein. For example, an acidically charged amino acid can substitute for a neutrally or basically charged amino acid, altering the mobility of the protein in an electric field and enabling separation of the two genetic forms of the protein by electrophoresis. The structure can also be altered by post-translational modification. For example, the ABO blood type is determined by whether sugar molecules are added to the protein at certain positions. This occurs after the protein has been synthesized, or post-translationally.

Protein polymorphisms are genotyped in several ways. Protein *alleles* (variants) can be detected by electrophoresis, which is separation by charge or size in an electrical field. However, several protein polymorphisms are genotyped by highly specific and sensitive immunological methods. The immunological methods depend on the use of an antibody molecule that recognizes the small difference in a protein (the *antigen*). The antibody can be generated by immunizing another individual or animal by injecting it with the foreign protein. The immune serum or purified antibody can then be used for the serological test.

In 1912 the Austrian immunologist, pathologist and Nobel laureate Karl Landsteiner discovered the first of the serological polymorphisms, the ABO blood polymorphism. Many people know their ABO genotype, which might, for example, be printed on one's military ID or driver's license. Our red cells carry the H protein, which if not glycosylated produces blood type O and otherwise can exist in the A or B form: three alleles (variants) at one protein locus. With three alleles, people can have up to four different ABO serological genotypes: O, A, B (all of which are homozygous genotypes) and AB (a heterozygous genotype). The OA and OB genotypes are also serologically genotyped as A and B. The ABO polymorphism is a very important one because common bacteria in our gut (*Escherichia coli*) make the A and B antigens. For this reason, an individual who does not carry these antigens herself will already have become naturally immunized against them. If a type O individual is transfused with type A, type B or type AB blood, he will immediately reject it. In contrast, type O (universal donor) blood can be transfused into individuals with the other genotypes, and a type AB individual is a universal recipient. To perform the test, a small amount of blood containing red cells is needed. Another important blood group antigen is the Rh antigen. People who have it are Rh positive. Many Rh-negative people have become immunized against the Rh antigen, and will reject Rh-positive blood.

DNA POLYMORPHISM

There are several types of DNA sequence variation, but two are most frequently important. The first is the *single nucleotide polymorphism* (SNP). An SNP is a single substitution in the DNA code of one nucleotide for another, for example the substitution of an A for a G at a certain position in the DNA sequence. Almost all SNPs are biallelic, meaning that there are two possible allelic forms (say allele 1 and allele 2) and three possible genotypes: 1–1, 1–2 and 2–2. For this reason, they are not as individually informative in many genetic analyses as the *short tandem repeat* type of DNA diversity. However, SNPs are perhaps the most important type of variation for two reasons. First, they are by far the most abundant, constituting approximately 99 percent of sequence variation. As mentioned, for the three billion nucleotides in a woman's genome, on average about 1 in 1000 is *heterozygous*, a different form having been inherited from her mother than from her father, so people contain about three million SNPs that are variable *within* their own genomes, that is from one chromosome to the second member of the chromosome pair. Second, SNPs are a type of variation that is very efficiently assayed, or *genotyped*. Commercially available genotyping arrays already enable up to five million genotypes to be simultaneously determined. This type of genotyping is illustrated in Figure 10.3. Notice that the DNA is analyzed on an array consisting of many millions of nano-sized features, representing the hundreds of thousands of SNPs being genotyped. The genotype signals are imaged, as shown on the right side of the figure, and the information then is interpreted and converted to actual genotypes using a map generated for the array after its manufacture and taking advantage of high-speed

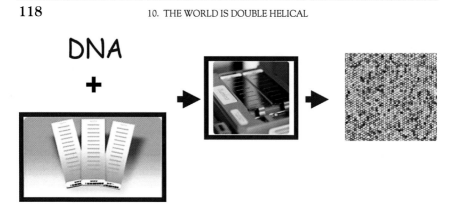

FIGURE 10.3   Array-based genotyping of millions of markers.

computing. Array-based genotyping of this type has only been available since the beginning of this century, but more than 99.9 percent of genotypes that have ever been generated have been performed in this way. The cost is very low, less than a penny a genotype.

Array-based genotyping of hundreds of thousands of markers is also available to the general public via commercial services that process the DNA from an easily collected saliva sample and offer rapid turnaround.

You may have noticed that genotyping is itself a method of DNA sequencing, where the sequencing effort is focused on places in the genome that are known to be genetically variable. Increasingly, the whole DNA sequence of a person is readable, especially by massively parallel DNA sequencing. When the human genome was first sequenced, a few hundred DNA fragments were read at a time. Nowadays, by sequencing about a billion small DNA fragments simultaneously, we can rapidly sequence an entire human genome, which is some three billion DNA bases in size, and with a sufficient level of redundancy to ensure accuracy. As I write in 2010, such sequencing is expensive and a genetics lab such as my own would only be able to sequence a few complete genomes a year. However, the $1000 genome may be available by the time this book is published and, more importantly, we will probably be able to completely sequence many genomes. Once the DNA bases that comprise the DNA code have been accurately transcribed into letters, A, G, C and T, they can be read just like any text and in paper or electronic form.

For many purposes a whole genome sequence is unnecessary. For example, for forensic identification the millions of genetic variants that are detected by whole genome sequencing represent a vast overkill. Only a few DNA markers of the right type are needed. Also, whole-genome sequencing demands a larger quantity of DNA than does analysis of a small panel of informative genetic markers. That small panel of markers sometimes is genotyped in many samples, so cost again becomes an issue. However, if a person's is sequenced once, for example at birth, the need for later genotyping of that individual would be largely eliminated throughout the rest of their lifetime.

Each type of genetic marker, from protein to DNA, and each method for "genotyping" and sequencing has its pitfalls and advantages. In research, there is constant innovation and diversification of methods, but when genotypes are used in medicine and in the courtroom the methods have to be stabilized and highly refined for quality, to minimize the possibility of error.

MEASURED ANCESTRY

More than nine-tenths of genetic variation is interindividual and only one-tenth is predictable on the basis of population of origin. This is a compelling argument against racism: on an objective basis one knows very little about a person merely from their race, skin color, hair or the shape of their eyes. However, because of the number of polymorphisms, the population differences that do exist translate into a very powerful tool for measuring the ancestral origins of an individual's genome, and even the ancestry of particular chromosomal regions.

To measure ancestry, geneticists use *ancestry informative markers* (AIMs). An AIM is a genetic locus specifically selected from among 22 million known genetic markers for its ability to distinguish human populations. Most AIMs can be rapidly and inexpensively genotyped. For the worldwide ancestry analysis shown in Figure 10.4, my colleagues Pei-Hong Shen and Colin Hodgkinson and I selected a relatively small panel of markers: only 186. Each was an SNP involving the substitution of a single DNA nucleotide for another and represented by only two allelic forms in the population. This was so that all of them could be genotyped inexpensively, accurately and rapidly on a genotyping array which also contained genetic markers used for other purposes. Each of the AIMs had a difference in frequency of at least ten-fold between populations; with all 186 together they constitute a powerful set for measuring ancestry.

What we see in the multicolored bar graph in Figure 10.4 are the ancestry scores for 1051 individuals representing 51 geographically distinct populations worldwide. This Human Genome Diversity Panel is in use by geneticists around the world. The individuals are grouped by their population of origin and the predominant ethnic factors are color coded. Already with this small number of AIMs – only a fraction of what is available – one can see that measured ancestry tracks with geography and population of origin in a remarkably powerful way. Reading

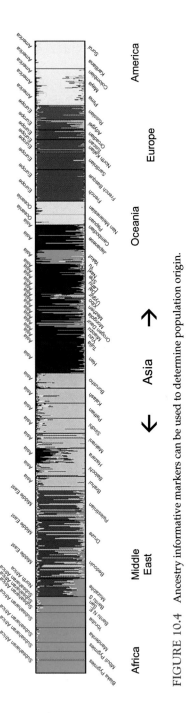

FIGURE 10.4   Ancestry informative markers can be used to determine population origin.

from left to right are Africans and African populations (blue). Next (red) are North Africans and Middle Eastern peoples. Note how the North Africans have some Middle Eastern ancestry component (red) but also some African ancestry (blue). Next are a series of Asian populations, with different components of Asian-specific factors. For example, both Chinese and Japanese are predominantly one of the Asian factors (dark blue). Continuing rightward (tan) are two populations from Oceania: Melanesians and Micronesians. European populations (red) and Native American populations (yellow) are at the right of the figure. This represents a seven-factor view of worldwide ancestry. Deeper views with more factors further subdivide some of the populations, for example Finns are closer to other Europeans than to other populations but nevertheless are distinct from other European populations. The ability to measure ancestry is a breakthrough in studies aimed at understanding the genetic origins of disease because many disease genes are common only in particular populations.

The first lesson from studying the ancestry of humans was that the vast majority of human genetic variation, more than 90 percent, is between individuals rather than population specific. Also, the population genetic variation does not fit perfectly against racial classification schemes. In this sense, critics of the use of "race" are correct. However, they are only correct in the sense that other descriptors are more precise, even in individuals who are admixed. Furthermore, despite the fact that the population-specific component of human diversity is relatively small, this component is functionally important. Certain functional polymorphisms are strongly tied to particular populations, or even found in these populations exclusively. For example, the hemoglobin S allele, the cause of sickle cell anemia, is common only in individuals of West African origin. The hemoglobin S allele is a classic single nucleotide variant; substitution of one DNA nucleotide is sufficient to substitute valine for isoleucine at the sixth amino acid of the hemoglobin beta protein. That one seemingly innocent change of one neutrally charged amino acid for another causes hemoglobin to form aggregates under conditions of low oxygen, leading to catastrophic damage to red blood cells. Studies have shown a single evolutionary origin for almost all hemoglobin S alleles and we even know why it spread and became highly abundant in West Africa: hemoglobin S confers resistance to malaria, and malaria is common there. Interestingly, West Africans and other malaria-afflicted populations have additional genetic variants that confer resistance against malaria and these are found both in beta globin and in other genes such as *G6PD*. In each case the variants are primarily population restricted.

Every population appears to carry variants that under some conditions confer an advantage and under other conditions lead to disease. Cystic fibrosis, a disease causing sinus infections and pneumonia, diarrhea and infertility, is abundant in northern Europeans and rarer in other populations. Owing to the discovery of the cause of cystic fibrosis by Francis Collins and other geneticists, we know today that 1 in 25 European Americans carries a single copy disease allele of the most common mutation, Delta F508, a three-nucleotide deletion that in turn deletes the amino acid phenylalanine at position 508 of the cystic fibrosis transmembrane conductance regulator (CFTR) protein. Also, it has been learned that the CFTR protein controls membrane chloride transport in many cells of the body, accounting for the diverse manifestations of the disease. The carriers of a single copy of Delta F508 or one of the many less common mutations that also cause cystic fibrosis are unaffected because the disease is recessive. However, some 30,000 Americans have inherited two copies, and therefore suffer from cystic fibrosis. Approximately 1 in 65 Hispanics and 1 in 90 Asians carry a cystic fibrosis mutation, but this translates to an even greater reduction in the frequency of the disease in those populations because of the recessive nature of cystic fibrosis. Also, different mutations are more common elsewhere in the world. Tracing the population associations of cystic fibrosis mutations may eventually clarify why the Delta F508 variant is so common in northern Europeans. The most likely explanation is that heterozygous carriers again have some advantage against infectious disease, possibly tuberculosis, cholera or typhoid fever, or even against lactose intolerance.

Parsing individual genetic ancestry increases the power of genetic studies via better targeting of studies, and prevents results being confounded by ancestry (for example, if patients were compared to controls of a different ancestry false conclusions might easily be reached). Furthermore, geneticists interested in the origins of human populations can reconstruct those relationships based on the *measured ancestry* of living descendants of founders of those populations. It can therefore be appreciated that the development of the capability to measure ancestry has been accelerated by scientists interested in disease gene mapping and in broad questions about population origins. Through these studies, geneticists have been able to confirm and greatly deepen our appreciation of the phenomenon of genetic individuality, as well as the genetic affinities of people who share common origins.

The Stochastic Brain
From DNA Blueprint to Behavior

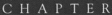
"Blind man breakin' out of a trance
Puts both his hands in the pockets of chance
Hopin' to find one circumstance
Of dignity."

Lyrics to 'Dignity', Bob Dylan

Clouds are not spheres, mountains are not cones, coastlines are not circles, and bark is not smooth, nor does lightning travel in a straight line.

Fractal Geometry of Nature, **Benoît B Mandelbrot**

Our Genes, Our Choices
DOI: 10.1016/B978-0-12-396952-1.00011-7

This chapter attempts to describe how the brain, with its astronomically complex interconnections and lifelong plasticity, builds itself using a restricted instruction set of some 24,000 genes, and perhaps an equally large number of regulatory RNA molecules. A crude statement of the problem of complexity is that the human brain contains about 100 billion cells that have to be "properly wired" into some 10^{15} (one million billion) connections. However, that vastly understates the intricacies inherent in the multineuronal networks necessary for memory, cognition, motor control, cognitive and sensory functions. For the time being, and with our present state of knowledge, it is most accurate to refrain from even putting a number on that number. Also, an effort such as this one to explain the process by which the brain develops can only sketch broad outlines and principles. Fortunately, other books have sketched or more completely developed the story of how the brain encodes emotion and cognitive functions, and for emotion I would point to Joseph Ledoux's *The Emotional Brain*, as a general introduction Mark Gazzaniga's *Nature's Mind*, and for the development of the brain and how the brain, through fundamental processes including neuronal plasticity builds our "self", Ledoux's *Synaptic Self*. For those making comparisons, my emphasis is on the unfolding of the genetic program.

SELF-ASSEMBLY

The brain is not built like a car. Cars do not self-assemble. The components of a car are assembled by the hands of robots and humans, who have programmed the robots. We can learn about complexity by studying mechanical constructions but cannot understand how biological entities develop and function merely by studying cars and similar inanimate constructions. The most amazing – but as we will see, explainable – aspect of the brain is that it is self-assembled.

It does not suffice to have the "right stuff": the stuff of life has to be in the right place and at the right time. Unlike components of a car, for example the steering wheel, the brain's molecules, cells and networks of cells have autonomous properties that enable them to recognize, respond and adapt to build, through a hierarchical series of interactions, the much more complex structure of the brain. These things are alive! Neither the steering wheel of your car nor the granule cell or Purkinje cell of your cerebellum has any understanding of the overall system in which it will play a part. Also, the steering wheel can be put aside until it is ready to be installed, but the Purkinje cell must be maintained and nurtured through interactions with other cells. However, because the Purkinje cell is itself a living entity it is also capable of responding in a myriad ways to its environment and neighboring cells with which it has synaptic interactions and metabolic cross-talk, and if not needed the Purkinje cell's disappearance (death) can be readily programmed. The steering wheel has only a limited "conversation" with the column to which it is attached and if the car is constructed without a steering wheel the steering column is oblivious. It will just sit there as if it is still doing its job. On the other hand, it is a rare instance in which a car is delivered that has two steering wheels. In development this type of error can easily happen because the functions guiding self-assembly are simpler and more readily confused if, for example, a chemical gradient is temporarily disrupted at the critical moment. The conversations between cerebellar granule cells and Purkinje cells are both nuanced and needful. They guide the development and function of both cells and enable the formation of hierarchically more complex levels which further drive the assembly of the brain towards its ultimate complex structure and capabilities. However, there is no third party standing by to say, "Ach! A car with two steering wheels – better fix that one."

Our complement of genes succeeds as a self-assembling instruction set only because it does not try to do everything at once. It is used to build molecules, then cells, and then networks of cells. All along during this process there are feedback loops of a complexity that is probably infinite, such that molecular expression and the differentiation, positioning and interaction of cells are in constant adjustment and refinement. Even in adult life there is no point at which a brain is static. The brain is constantly adapting at molecular, cellular and network levels.

As will be seen, the brain's DNA blueprint succeeds because each genetic element in the instruction set is designed so that it can be subject to subtle regulation, including alternative molecular forms. The genes talk to each other and guide each other. The gene products (proteins and RNA molecules) interact in what has come to be known as an "interactome", the new and often useful fad being to place the suffix "ome" behind any concept, and form a new field of scientific inquiry (a new piece of the "inquirome"?). Unlike the tools that humans have traditionally manufactured, brain development involves transformation from one developmental stage towards the next, and it is self-organizing. Like the molecules, the neurons also talk to each other and guide each other, but their interactions are far more complex and nuanced. Scientists have physical models for many molecule–molecule interactions but are far from being able to model cell–cell interactions with anything approaching that accuracy and precision, as will soon be seen when we discuss the neuronal "connectome". The development and maintenance of neural organization is guided by function in accord with the "use it or lose it" principle and this depends on the ability of neurons to interact, and ultimately translate their interactions into molecular changes that may reinforce connections, weaken them or even signal an individual neuron that it is time for that neuron to die. Such "programmed cell death", or apoptosis, involves the activation of a particular molecular cascade, with many components.

Governing the whole process of brain development and its ongoing plasticity is a mysterious and ultimately fruitful interplay between our innate and individual complement of genetic variation, and stochastic (random) processes that enable infinite divergence in the neural networks of our brains. This stochasticity is the raw stuff that neuronal developmental processes mold into unique circuits. In this way, the story of the development of a brain, or even a local neuronal network, never unfolds in exactly the same way, no matter how many times the story is told. At an intrinsic genetic level, most humans, except for identical (clonal) twins, are unique, but during development we all become individual and, in humans, that individuality is over time amplified by the choices we make, which soon have the effect of shaping our brains in ways that we have chosen, whether wisely or not.

CELL ASSEMBLY

As mentioned a little earlier, neurons are complex entities, and to understand the self-assembly of a brain we have to first understand the self-assembly of neurons. The building blocks of cells are proteins, structural and regulatory RNAs (ribonucleic acids), and small molecules, such as the lipids in cell membranes, that proteins build or properly position, compartmentalize and modify.

The "central dogma of molecular biology" is DNA $\rightarrow$ RNA $\rightarrow$ protein. DNA is transcribed to messenger RNA by an enzyme known as RNA polymerase. RNA is processed and then translated into protein by cell machinery known as the ribosome. The proteins in turn fulfill myriad roles, forming the structure of the cell and scaffolding, barriers and transport highways for other proteins and small molecules, catalyzing chemical reactions, transporting other molecules, regulating genes and each other, signaling distant cells and serving as receptors for such signals.

This "central dogma" has been an extremely fruitful and powerful explanatory model for the capacity of DNA to direct the molecular construction and function of cells. However, keep in mind that there is always a good chance that someone's karma will run over your dogma, and over time important exceptions to the central dogma have been discovered.

Exceptions to the central dogma include the ability of some RNAs to be reverse transcribed into DNA (RNA $\rightarrow$ DNA) and the discovery of RNA molecules with intrinsic structural, regulatory and catalytic activities. It has recently been appreciated that there are probably at least as many regulatory RNA molecules as there are protein-coding ones. These new functions for RNA extend the array of potential mechanisms by which DNA can ultimately direct cellular structure and function. In addition, alternative processed forms of RNAs can enable a "single gene" to encode a multitude of protein forms, even before post-translational modifications of proteins, including additions of phosphate and carbohydrate groups, altering their structure and functional properties.

Ultimately, the DNA directly or indirectly encodes an order of magnitude more than 24,000 large molecules, but does so in a fashion that is open to specific regulatory control. Thus, there is a great range of molecular plasticity for the cell to exploit during development and differentiation, as it faces different functional challenges. The epigenetic plasticity of the genome will be taken up further later, and in the more global (rather than cellular) context of gene by environment interaction.

THE PROTEIN INTERACTOME: SIX DEGREES OF MOLECULAR SEPARATION

The proteins and RNA molecules encoded by genes work together to build cell structures and functional molecular networks. Scientists of several related disciplines – molecular cell biology, biochemistry, neurobiology, endocrinology and genetics – have now achieved a significant, but partial, understanding of the molecular networks that govern the function of cells. Examples of some of the better understood pathways include regulation of the cell cycle and DNA synthesis, and cell energy generation including oxidative phosphorylation and glycolysis (the process by which sugars are metabolized to regenerate ATP, providing energy for many other functions). Increasingly, it is understood that many of these systems of interacting proteins are better understood as networks, rather than as linear pathways. A major task of genetics and molecular cell biology is to fill in the missing pieces of molecular networks and indeed there are still many thousands of genes whose functions are unknown and that must represent such missing pieces, or new networks.

Even proteins of unknown function are amenable to many types of analysis. The levels of the RNAs that encode these proteins have been accurately measured and correlated with each other, cellular functions and genetic variation at different times of development, in different tissues and under different conditions. The physical interactions of the proteins themselves can be directly measured and they can be colocalized within cellular compartments. The unknown players are being found to connect to known molecular networks, where they play new, and sometimes more cell-specific roles. The complexity of the cell's molecular networks is immediately obvious to any biochemistry student forced to memorize the glycolytic pathway (but why should someone have to be forced?). However, the new RNA and protein technologies just mentioned, and systematic analyses of the published results on interactions, have enabled the concept of the protein interactome to be taken to the global cellular level. For those who don't believe in reductionism, accommodate yourselves to the fact that after all our studies of the protein interactome, where any protein can be related to the function of any other through six degrees of separation, what we end up with is … Kevin Bacon (Figure 11.1).

If the manufacturing process is perfected, an automobile can be built from perfectly machined parts such that it works each and every time, and one car of the same make and color is initially indistinguishable from the next. Having come this far, to the level of the protein interactome, let us step back and summarize why Kevin Bacon's brain cannot be built that way. There are three answers to this question. The first is that the developing brain is a living, functioning thing, not a piece of incomplete machinery on an assembly line. Therefore, it must develop from one viable stage to the next, as the developing body partially recapitulates its phylogenetic origins. Second, nothing as complex as a brain could be specified from an instruction set of only 24,000 genes and perhaps an equal number of non-coding RNAs. The amazing fact is that brain development can be programmed from such an instruction set. Third, the plasticity inherent in a construction by a developmental process, rather than a piece-by-piece assembly, has given living organisms an enormous advantage in being able to adapt to different conditions

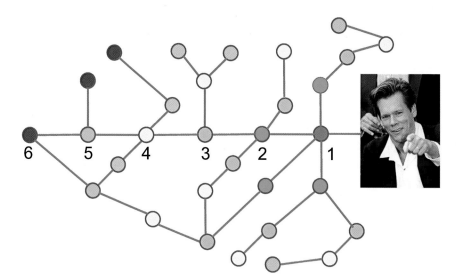

FIGURE 11.1 The protein interactome: six degrees of molecular separation.

present during development, and indeed our brains continue to exploit this plasticity throughout adult life.

STOCHASTICITY

How do coherent, adaptive behaviors emerge out of the complexity of random molecular chaos? Turning the question around, how does the chaotic, unpredictable setting and unfolding of our genetic program lead to individuality and choice? Is there something about our brains and genomes such that when we put "both hands in the pockets of chance", free will appears as an emergent property of the human condition?

Plasticity, which is the capacity for change, and random variation (also known as stochasticity, from the Greek *stokhos* – a guess) are hallmarks of the human brain. The fingers of identical twins closely resemble each other in length and shape. One twin's fingers predict whether the other's are long or short or webbed at the base. However, twins' fingerprints are easily distinguished one from the other. If your evil twin wants to frame you for a crime, he should leave his DNA, not his fingerprints. The pattern of fingerprints is subject to general rules, but the details of the position and number of every ridge and whorl are not programmed. The same goes for other external features of a person. Hairiness, hair color and general distributions and whorled patterns are inherited but the position of each hair on one's head and body is random. Similarly, brain size and major characteristics of personality, cognition and underlying neural mechanisms are strongly inherited; however, we are each individual – as individual as snowflakes.

SNOWFLAKES AND NEURONS

Within a snowstorm are billions of unique snowflakes and within our brains are billions of neurons, each with a unique pattern of synaptic arborization (tree-like branching). The general form and behavior of snowflakes and neurons obey certain general rules. Owing to the packing angles of water molecules, snowflakes have a general hexagonal symmetry and thus could never be mistaken for crystals formed from certain other molecules, for example the needle-like crystals of uric acid that plague gout patients. Pyramidal neurons of the cerebral cortex which send yard-long axons to the ventral root ganglion (neuronal cell mass) of the spinal cord all resemble each other much more strongly than they do other neurons or other cells of the body, and again their group coherence is due to the action of molecules that comprise them.

There the similarity ends. Pyramidal neurons, and other neurons, are each comprised of a complex, constantly shifting and reactive network of molecules adapted to work together in pathways and networks that function within the cell and across cells. Within the cell these build structures: nuclei, nucleoli, mitochondria, ribosomes, vesicles, Golgi apparatuses, which themselves vary in position of shape from one cell to the next. A snowflake might be compressed against its neighbors to be physically broken or fused, but it experiences only a shadow of the interactions of a neuron with its neighboring neurons and with the multitude of other cells that compose its environment. The snowflake does not have the "inner life" of a neuron, and thus by comparison its repertoire of possible responses is extremely limited. Snowflakes can melt and neurons can die, but neurons have a much wider range of reaction, and potential for random variation to alter that. Each neuron is not merely pressed against its neighbors but is dynamically interconnected, communicating electrochemically and metabolically with nearby cells and forming vastly complex networks that in the end enable the brain to produce all the outputs of the mind. A neuron is constantly being structurally and biochemically shaped by its neighbors but is also shaping them, and networks of neurons are shaping other networks, and the cells within them.

BRAIN GENES: CASCADE EFFECTS, CHAOTIC EFFECTS AND GREAT ATTRACTORS

Complex systems are not necessarily self-correcting. Complexity can instead make systems more vulnerable. One change, or defect, can lead to a cascade of effects. For example, with all its complex components, a space shuttle may fail because of a single glitch in a computer code. Or perhaps someone hits the wrong button (famous last words on a certain flight recorder: "Hey, what's this button do?"). Recognizing that fact, backup systems are built into a space shuttle and our brain also has backup systems and mechanisms of compensation. Nevertheless, despite the backup systems, some errors lead to disaster.

To make matters better, and understand how the brain can work, it is necessary to descend into chaos theory. Chaos theory is closely associated with the French–American mathematician Benoît Mandelbrot, the father of fractal geometry. This theory defines some key features of the behavior of complex systems. An important feature of natural systems is their irregularity both at the macrolevel and at progressively finer degrees of magnification, as seen in the shapes of leaves and blood vessels, and the neurons' complex branchings, internal structures and biochemistry.

The complexity visible at the macrolevel is reproduced at the microlevel – a phenomenon known as scale invariance – and leads to essentially infinite complexity. Thus, the irregularity of a shoreline observable from outer space is also observable at the microscopic level. In his paper "How long is the coast of Britain? Statistical self-similarity and fractional dimension", Mandelbrot showed that Britain's coastline is infinite in length, if one uses an infinitely small measuring ruler. This infinite complexity of natural systems such as the brain is closely related to the occurrence of chaotic behavior in such systems. Chaotic behavior, which is often called the "butterfly effect", is a property of certain types of change within certain types of dynamic system.

As Frank Herbert observed in *Dune*, "A beginning is a very delicate time". A very small change in the "starting" state of the system can lead to a very large difference in the outcome. One thing leads to another. The result can be that even though all the component interactions within the system are deterministic the system appears to behave randomly, because the very small changes in initial starting state were essentially unmeasurable. The seriousness of the effects of chaos for understanding how the brain (which has billions of neurons) functions is emphasized by the fact that the mathematician Poincaré first discovered chaos in the 1880s when studying the three-body problem!

BRAIN ASSEMBLY

The locations, arborization structures and biochemistries of billions of neurons cannot be specified, but their development can be programmed through a genome-guided developmental process. The brain programs itself, and it does so by a process of neural development by cellular interaction, beginning from the earliest moments in development when a cell that happens to be on the outside of a cell mass experiences a different environment than a cell on the inside of the same cell mass. These small starting differences lead to a cascade of changes. Also, and because of the unprogrammed molecular and structural complexity of cells, even two cells that are positioned so that they receive nearly identical influences will begin to diverge in subtle ways.

Immature neurons derived from an embryonic structure known as the neural crest migrate to their final positions along a scaffolding of radial glial cells – actually crawling along these glial cells for long distances. Neurons born later migrate past those that have come before, occupying more remote territories, and the cells that are born later are already intrinsically different, leading to differences in the connections they "want" to form. After cells reach their final positions they begin to grow axons and dendrites to make those connections, the growth being steered to the appropriate targets by contact with other cells and by combinations of adhesion molecules that are fixed in place and chemotactic molecular gradients of small molecules diffusing in the space between cells, all of which the amoeba-like neuronal growth cone encounters as it explores the environment. The growth cone of an axon often takes a route pioneered by the axon of another neuron, leading to the formation of neuronal fiber tracts. However, after neurons migrate into place and complete the process of forming axonal and dendritic connections with other cells, as many as half of them die. This massive cell death is thought to be due to competition between neurons for the attention of target neurons, including neurotrophic factors that are necessary for survival and that are secreted by target neurons. In many parts of the brain a dense network of synaptic connections is then formed which will later be refined by the elimination of synapses, in a process guided by use. Developmental sequence and basins of attraction thus tame a chaotic landscape of unpredictability. The brain is not created all at once or in a random way, but emerges in a developmental sequence, one moment in development setting the stage for the next.

In adults, neuronal plasticity, and even neurogenesis – the generation of new neurons – is not finished, and adult neuronal plasticity is important in many ways. As elegantly detailed by Jay Giedde and colleagues at the National Institute of Mental Health (NIMH), and by others, myelination increases dramatically after birth, and also there is progressive synaptic pruning through young adulthood. This is seen in Figure 11.2, where through brain imaging the gray matter density can be seen to "cool down" from red and yellow colors to green and blue in these heat-mapped images. Adult neurogenesis was first well appreciated in songbirds. Birds such as the canary that are capable of learning new songs require new neurons. However, it has recently been appreciated that adult neurogenesis is also important in humans and other mammals, and in at least in two areas of the brain, the olfactory bulb and hippocampus. This is illustrated in Figure 11.3, where it can be seen that in both cases the new neurons have to migrate considerable distances to find the proper locations, and then somehow integrate with neuronal networks. As shown by Rene Hen and his colleagues at Columbia University, adult neurogenesis in the hippocampus is crucial for recovery from depression, at least after treatment with the widely used selective serotonin reuptake inhibitor (SSRI) antidepressant drugs. This was proven in a very specific way: by irradiating the small region of the brain that has the neural stem cells that repopulate the hippocampus, Hen was able to block recovery in a rat model of depression.

In its staging, the developmental sequence of the brain resembles an expedition to climb a great mountain. The climbers do not scale the peak

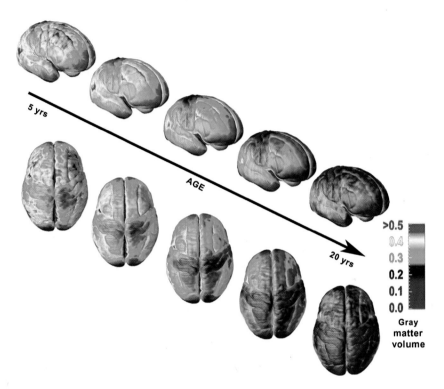

FIGURE 11.2 Gray matter change from childhood to young adulthood. *(Source: Gogtay et al., Proceedings of the National Academy of Sciences of the USA, 2004)*

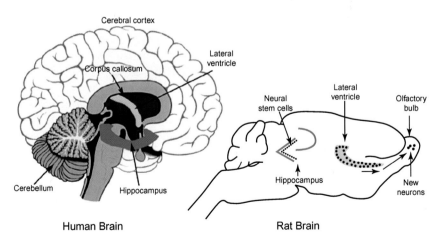

FIGURE 11.3 Neuronal stem cells and adult neurogenesis in the hippocampus (left) and olfactory bulb (right). *(Source: Crews and Nixon)*

all at once, but move from base camp to progressively higher waystations on the mountain, which they establish themselves. Some are helped more by other climbers. Some turn back on the way, or even die. In the development of the human brain, more than half the cells do not survive.

USE IT OR LOSE IT

As said by neuroscientist Carla Shatz, a major self-organizing principle of the brain is "use it or lose it". The preservation of synapses depends on their excitation. Synapses and circuits that are used are strengthened and those unused are lost. Early in life, we pass through stages of great plasticity when language, music and motor skills are most readily acquired. This is dramatically seen in the visual cortex, where the neuroscientists Hubel and Wiesel famously showed that there was a narrow window of time – only about a week – when the occipital cortex must receive input to properly develop. If one side of the visual cortex is deprived, it is difficult to reverse the imbalance (amblyopia) that develops. Although amblyopia is called "lazy eye" it is actually "lazy brain".

THE FRACTAL NATURE OF INDIVIDUAL NEURONS

The larger structures of the brain are programmed, but not the locations and interconnections of individual neurons or the fine structure and function of networks in the developed brain. In this regard, the human brain is very different than the brain of simpler organisms such as the nematode worm *Caenorhabditis elegans*. The identity and position of each of the 302 neurons of *C. elegans* can be forecast in advance. Much less can we predict the shapes of the extensions of the neurons in our brain, the synapses with other cells to form neural circuits, the locations of neurotransmitter-activated channels within a synapse, or the positioning and exact functional efficiency of individual secondary signaling proteins within the cytoplasm. Indeed, each of these is constantly changing in dynamic response to functional demands. However, each of these phenomena is regulated by chemical and electrical gradients, and by interactions with other cells, and these interactions unfold in a temporal sequence. Rules shape the randomness.

The shape of the maple leaf is fractal but each maple leaf looks more or less the same because of gradients of chemical and cellular interaction during development. If a gene programming the pattern of the maple leaf should be altered, the jagged form of the maple leaf might be smoothed to more closely resemble the leaf of some other tree. Similarly, there are many different types of neurons, several of which have highly distinctive shapes to enable them to carry out their functions. The variety and complexity of neuronal shapes were first appreciated when the legendary Spanish neuroscientist Santiago Ramon y Cajal exploited the silver stain (the "black reaction") invented by Golgi, who shared the Nobel prize with him. Curiously, only one or a few neurons will intensely pick up the silver stain, allowing the forms of these neurons to be picked out from the surrounding cell mass. As Floyd Bloom said, "The gain in brain is mainly in the stain." When I was a postdoctoral fellow I helped my mentor Carl Merril adapt the Golgi stain for the staining of proteins and nucleic acids in electrophoretic gels, and it very quickly became a widely used, ultrasensitive and quantitative method – I am sure that many more gels than brains have been silver-stained. But in the brain, where the silver stain detects cells, the structure that is revealed is far more interesting, The variety of neuronal cell types and configurations detected with the Golgi stain – each of which is an individual three-dimensional fractal structure – is illustrated in the two-dimensional line drawings in Figure 11.4.

HIGHER ORDER BRAIN STRUCTURE AND RANDOMNESS

As compared to a leaf's or neuron's individual patterns, the higher order patterns of the brain, dubbed the connectome, are much more complicated. The technicolor brain slice images in Figures 11.5 and 11.6, which are called "brainbow images" illustrate the complexity of neuronal layering and interaction at different levels of the brain's architecture. This

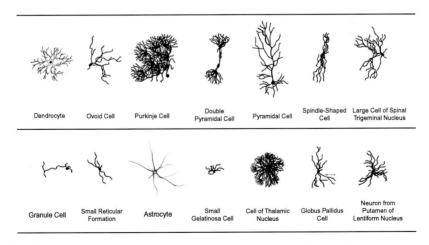

FIGURE 11.4   Illustrations of various types of neurons found in the CNS. (*Adapted from Ramon y Cajal, following 'DNA repair, mitochondria, and neurodegeneration', Weissman et al., Neuroscience, 2007, 145:4, pp 1318–1329*).

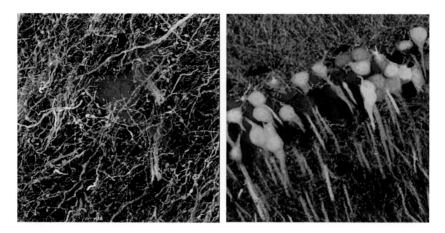

FIGURE 11.5 Brainbow images of a neuron (left) and neurons (right). *(Source: Jeffrey Lichtman).*

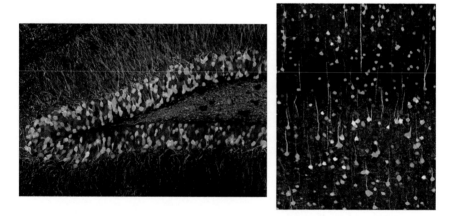

FIGURE 11.6 Brainbow images of neuronal fields. *(Source: Jeffrey Lichtman).*

revolution in simultaneously imaging all the neurons in a brain region is the product of the imagination and technical skill of Jeffrey Lichtman and his colleagues and may well lead to another Nobel prize, particularly if sufficient scanning and computational resources can ever be brought to bear to enable the deciphering of the almost impossibly complex connectomes of even very small regions of the brain. These are not simply painted images of cells. Through a genetic trick, the cells paint themselves. A cassette of fluorescent protein genes is inserted and in different cells these genes are rearranged in different ways. The fluorescent proteins are of several colors, and as they are simultaneously expressed, they generate a palette of different colors so that nearby cells and their processes can be individually distinguished.

At the first level (Figure 11.5) is an individual neuron fluorescing red and embedded in a field of processes from neurons emitting other colors. It can be seen that these images enable reconstruction of not only the complex forms of individual neurons, but also their interconnections. Thus far, the technology has been most successfully applied to supposedly simple systems, in particular the neuromuscular junction, where enormous and functionally meaningful variation has nevertheless been observed in the way neuronal connections are formed – variations that are beyond the ability of genes to specifically control.

The brainbow images of neuronal fields in Figure 11.6 hint at higher levels, and different types, of architectures formed by neurons in three-dimensional fields of cells. The interactions of multiple neurons in a volume of brain are suggested by the cross-section on the right, where there is a layer of neuronal cell bodies, and the axons have the same predominant (downward) orientation.

Images such as this are inadequate to the task of defining the connectome in more than general terms, but are already sufficient to confirm that the brain has infinite randomness, as well as an architectural framework to guide, or constrain, the randomness.

THE STOCHASTIC BASIS OF BEHAVIORAL INTELLIGENCE: GUIDED RANDOMNESS

By now it should have become obvious that it is fundamentally wrong to think of the stochastic brain as a random system in the sense of the movements and collisions of dust motes. The brain is a system with an evolved "intelligence" that guides the behavior of its individual particles. A flock of blackbirds wheeling in the sky or a school of fish moving in unison has randomness but also cohesion. Such biological assemblages wheel and turn, transform constantly in shape and extent and make the most fantastical patterns beyond the comprehension, and certainly the planning, of the individuals who comprise them. However, group cohesion and patterning are driven by the dim awareness of each individual element – bird or fish or neuron.

Like a neuron, each element of the flock has extensive homeostatic and repair capabilities. The individual bird instinctively feeds and reproduces so that after it has died there is another to replace it. During breeding season individual birds pair and nest, "replenishing the ranks". Yet each fall and as days become short the flock reassembles and behaves as it has for many years past. Each winter evening in my part of Maryland a living river of hundreds of thousands of blackbirds flies from food sources such as harvested fields to roosting trees. These birds behave in this complex way because of genetically encoded general programs of behavior. Other winter-resident species, such as chickadees, bluejays, wrens, barred owls, bluebirds and pileated woodpeckers, have their own distinct behavioral patterns.

Each evening, the people who live beneath the trees favored by the blackbirds await the arrival of their fine feathered friends. Such flocks avoid woods such as mine, which are too wild and jungle-like. Arriving at their park-like destination, the flying horde coalesces into a cloud, known as a murmuration, that wheels, turns and morphs into fantastical shapes until it finally settles into the trees where it makes a cacophony. The birds pack as close together as Hitchcock's on the jungle gym of the little school where Annie taught. Bicycling home each autumn I hear taped bird distress calls from one troubled house. Assuming that the homeowner is attempting to drive off blackbirds, I somewhat sympathize. However, the group mind of the flock is stubborn. The flock has a mind of its own, and while the behavior of any of its individual elements would never be predictable, the landowner can predict with a very high degree of certainty that although he has not seen them for six months the birds will be back.

Like such murmurations of birds, our brains are enormously complex and unpredictable at the microlevel, and yet are more predictable at the macrolevel. Major brain structures and types of organization, for example the cortical layers, are preserved in most of us. Our brain circuits of reward, emotion, implicit memory, explicit memory, vestibular, autonomic, motor, sensory, auditory and olfactory perform as they were selected to perform by millions of years of selective pressure and often they display remarkable resilience in finding their way back to particular patterns of interaction and function, just as they always have. The blueprint is strong and part of the story of this book is the effect of gene variations within that larger, genetically defined blueprint. However, the exact patterns formed from that blueprint are not predictable in detail.

At the DNA and biochemical level there is submicrosopic randomness (stochasticity) that individualizes all neurons one from the other, but really this is only the molecular signature of the functional individuality of each cell. In this regard, humans are unlike digital computing devices that rely on very large numbers of structurally and functionally identical components that can be linked together in different ways. The mind's computational complexity is not merely a matter of its being composed of analog units capable of graded responses – computers can also be constructed from such components. We are also unique because of the variability in our parts. No two people can be alike because no two of our neurons are even alike. However, our individuality does not consist of randomness. Randomness is only the raw material from which a new ordered and individual neuronal complexity is extracted, and the rules set down in our genomes guide that process to destinations that are unique, but human.

RULES THAT GUIDE THE CHAOS OF BRAIN DEVELOPMENT AND EVOLUTION

There are rules for the development of a brain, and genes help define some of the rules, enabling our brains to form completely different structures than a fruitfly's, although our brain and the fruitfly's both develop in a similar general context of physics and biochemistry. Within the chaos are "great attractors", both genetic and physical, that lead to confluences of outcomes in the structures and functional capabilities of brains. The fruitfly brain has a mushroom body; ours does not. Some of the constraints on structure are physical and some are accidents of evolutionary precedent. It might be handy, as Dr. Evil might have wished, for a shark to have evolved a laser beam with which to kill its prey, but the physical requirements are high and the evolutionary precedents available for adaptation or exaptation were non-existent. Similarly, it would be advantageous for a person to be able to read others' minds, so why has that never evolved? Obviously because it was not so easy to get to that adaptive peak, even though with powerful technologies we may eventually be able to reach it.

When the confluence of physics and genetics has enabled a complex structure such as a brain to be built, that structure can be easily disrupted if some piece is missing. It is easier to disrupt or destroy a thing than to build one. Lacking any of a myriad necessary nutrients and aspects of physical environment, the brain cannot develop properly. If a key gene is missing, the result can be a gross failure in the programming and in the resulting structure of the brain. For example, lissencephaly (an inherited disease marked by a smooth cortical surface) can result from the action of single gene defects; so can microcephaly (small brain). Human brains have important similarities to the brains of other mammals, including non-human primates. The brains of all these mammalian species develop the same generally similar large structures such as hippocampus and cerebellum, and these structures serve the same functions. However, certain key genes have recently been identified that are responsible for some of the large-scale functional differences between humans and even our nearest relatives. One dream of deep sequencing the genome of man's closest relatives is to identify gene differences that are critical to human language and creativity. By sequencing ancient DNA, Svante Paabo at the Max Planck Institute has discovered that some Neanderthals, extinct relatives of man, cross-bred with ancestral humans such that some Neanderthal DNA survives in the human genome. But what are the genetic differences that may have limited Neanderthals and led to their ultimate displacement? Profound differences in language and cognitive abilities are ultimately attributable to these gene-programmed differences. Neanderthal's neurons could not simply do whatever they wanted to do, or what was needed when faced by competition with *Homo sapiens*.

GENETIC INDIVIDUALITY, CHAOS
AND SENSE OF SELF

Most people have a sense of self that extends all the way back to the earliest years of childhood. We feel that we are the "same person" now that we were then. Perhaps we even are that person. Old man (me!) looks in the mirror and instead of seeing himself as he is, sees the boy who became the man. We are a disorderly archive of memories, both explicit (conscious) and implicit (unconscious) of all our earlier versions. We may well have the same temperament and abilities (good with numbers, a genius at music, irresistible to the opposite sex – if this is what one sees in the mirror could it even be narcissism?), but in certain profound ways we are not the person we were. Each day we live, neurons and neuronal networks change, many neurons die and some are replaced by other cells in a phenomenon known as adult neurogenesis. While we sleep and dream our brains reshape themselves. We start each day anew.

In the ultimate abdication of responsibility, how can the brain-person who awoke this morning be held to account for actions (positive or negative) of its predecessor? Was that not "someone else"? No. There are two answers. First, the brain adaptively and randomly changes, but also fights for homeostasis, is a process that lasts as long as we live. Second, our "self" is not a thing but a system of numberless interacting things. Much is transformed, but self endures. The leopard cannot change its spots, but more to the point the leopard changes from day to day but never into a giraffe. The flock of blackbirds discussed earlier continually replaces its individual elements but preserves continuity of behavior through space and across seasons. However, it is a misconception to imagine that physical continuity is even a requirement for conservation of self. Many "things", such as ocean waves, require no continuity of physical elements as they translate through space and time. The great wave that crashed against the headland was the rising swell that could have been observed a mile offshore, although it brought with it none of the water molecules across that space. Indeed, like ocean waves we are complex waveforms translating through space–time, and subtly or more dramatically altering in form and function as we move forward. One of the true tragedies of severe brain diseases, and death, is this loss of self as the complex waveform dissipates, even as some effects of our being do translate forward far into the future and affect other lives.

A human self is like a great hurricane born as a confluence of bad weather south of the Canaries, organizing and gaining strength as it moves into the Eastern Caribbean, thundering across leagues of ocean, and making landfall in dry South Texas, where it loses all distinct being. Throughout its "life" the hurricane structure and behavior are defined by physical laws, which is why hurricanes resemble each other as a class and why weather forecasters can roughly predict their strength and course. Their exact places and times of birth and courses are impossible to predict. This lack of predictability does not, however, make a hurricane free. It is unaware of its past and its possible futures and does not select between them; it goes where the steering currents take it. Like the waveform of a hurricane, our human brains develop and translate through space–time in accordance with physical laws, but in addition we are controlled and informed by genetic and cultural instruction sets. Hurricanes exceed in size and power all living entities on Earth but generation after generation they remain the same. The hurricane has no progeny and leaves no genetic or cultural legacy to the next generation. It sings a song of wind and wave, but the song remains the same. It never learns unconsciously or consciously to alter its behavior by observing itself or others of its kind, for example to learn to avoid the mountainous island of Hispaniola. If there were no human alive to hear its wind song, the hurricane and its winds would have existed, but it is due to conscious observation by people that there is a lore and science of the hurricane, and this lore grows from generation to generation within a human framework outside a hurricane's ken.

It is a mystery how one could define the moment of beginning and ending for any complex entity, but it is a special puzzle to demarcate the temporal boundaries of the human self. It is the thesis of this book that neuroscience and neurogenetics yield the conclusion that humans are free, self-determining entities, deserving of respect. If this is true, then it is also crucial to define our beginnings and endings, and there are endless arguments on this point. If we become humans worthy of respect as free entities, when precisely do we deserve this reward, and responsibility? Is it the moment when haploid eggs and sperm, already genetically unique due to recombination and mutation, are formed? Does it happen at conception when sperm fertilizes egg and we become a single-celled, diploid life form? Or is it the eight-cell stage, the blastula, the gastrula, the quickening, birth or perhaps when a child becomes capable of perceiving his own folly? If I thought I knew, I would say. However, what is clear is that the puzzles of beginnings and endings require more careful exploration, and are probably beyond the bounds of this book, which I would say only sharpens the need for better answers. That discussion will also force us to confront the evolutionary beginnings and endings of humans and other life, and to evaluate the workings of the minds of other species (which is difficult because we find it difficult to imagine how they may think), and cybernetic systems, as touched upon in Chapter 19 in commentary on the HAL 9000 computer.

Categorizing complex systems as alive versus unalive, free versus unfree, or conscious versus unconscious, is an exercise inherently fraught with error. We can state the precise moment when a cluster of thunderstorms becomes a tropical storm and a tropical storm a hurricane only because people have chosen to arbitrarily define this on the basis of sustained wind velocity. At some point the victim of Alzheimer's disease loses touch with enough parts of their own self that the body and brain that go forward are no longer the former self, and at some point there is no mind to make free choices. Long before that point there may have been sufficient impairment to justify legal guardianship, but here I am addressing a deeper and more difficult point. However, even as we contemplate beginnings and endings, we are also conscious that brain diseases can impair continuity of self in limited and sometimes reversible ways. Like a troubled electrical system, power can fade in and out, and our consciousness and will exist only as physical manifestations – there is nothing ethereal about them. Turn off the power and the machine stops.

In an amazing demonstration of the fact that the brain is the person, any of us may suffer temporary amnesia or loss of a motor or sensory or language ability, and recover. As discussed earlier in much greater depth, psychiatric disorders can distort our sense of self and locus of control. Addictions can make people become unlike their "true selves" in certain ways. The addict may steal, lie, crave and suffer misery as a result of their addiction and to a limited extent we can now help addicts and restore the addicted brain, but much better tools need to be developed beyond just the strategy of prevention. The depths of depression, the whirlwind chaos of mania and the bizarreness of psychosis all disrupt the mind and are reversible, often spontaneously and often with the help of pharmacological or other intervention – even electroconvulsive therapy or electrical stimulation of particular regions of brain. Ultimately, as in the deep stages of dementia or in the death of the body that is the destiny for all of us, the thread of continuity between what we were and the diminished thing that we become is frayed. At the end of life, and like the thread of fate woven by the Norns in Norse mythology, it will ultimately be broken. As observed in Citizen Kane, aging is the one disease for which one does not want the cure, but the end of life, and the aging process, inexorably approaches: no one gets out alive. However, if we are part of a continuing society of people, or perhaps a component of a community of other intelligences, something lives on of what we did and the spirit that guided us. The blackbird dies, but the flock lives. Also, like the Neanderthals who are now all long dead, we may transmit a genetic legacy to new generations who carry within themselves part of our DNA code, our "selfish genes" that through their selective advantage may achieve a type of immortality. Also, and again as impeccably told by Richard Dawkins, our memes may survive and these may be transmitted both horizontally in society and as a legacy to generations that follow.

Reintroducing Genes
and Behavior

"One for sorrow, two for joy. Three for a girl and four for a boy. Five for silver, six for gold. Seven for a secret, never to be told ..."

One for Sorrow, **Traditional Childrens' Nursery Rhyme**

Half way through the journey of this book, this is what has happened so far. We started with a gene – *HTR2B* – "for" behavior. A severe and common genetic variation in this neurotransmitter receptor was partially predictive of severe impulsivity, and posed a challenge to the conception of free will. Next, we brooded over the measurement of behavior and the reification of some patterns of behavior into psychiatric diseases. We thrashed out the politics of behavior and behavior genetics, including the gene–eugenics connection, and the genethics of psychiatric genetics, and showed the hollowness and inconstancy of conceptions of moral ethics in which people are treated as free even though in our hearts and brains we never acknowledge their autonomy. We visited the bestiary of psychiatric diseases that distort an individual's ability to choose. This was followed

Our Genes, Our Choices
DOI: 10.1016/B978-0-12-396952-1.00012-9

by a step-by-step analysis of how a genetic blueprint can build a brain, and how intrinsic to that process is stochasticity that amplifies the neurogenetic individuality with which we began, and yet has rules that, by and large, guide the self-assembly of the brain successfully, using the dim awareness of its individual elements. So that is a brief *in medias res* summary that might be useful if, as is my practice, you picked up this book and randomly happened to begin reading this page located just past the midpoint.

We return to the level of the DNA blueprint, and its role in individuality and behavioral prediction. To understand the potential and the reality of present-day behavioral genetics, we will start with broad technological capabilities and exaggerated commercial claims for genetic prediction, and then follow with an explanation of the classical process by which functional genetic variants have been identified, and tied to behavior and disease.

What are the limits of genetically based behavioral prediction? Are some secrets of the mind forever inaccessible to genetic and physiological probing? In the not-so-futuristic movie *Gattaca*, the protagonist's ambitions are threatened by his "inferior" genome. His struggle is as much against a presumption of inferiority as with innate handicaps. In this world of the future the genomes of the well-to-do are engineered to eliminate glitches that lead to diseases and to enhance intelligence, stamina and special talents. The engineering is dramatized by a pianist with six digits on each hand (a genetically engineered polydactyly). In this world of the future, from birth, people are tracked by their DNA and the assessment of a potential new girlfriend includes surreptitious DNA analysis, on the basis of which her talents or suitability are judged. This is judging a person by the genome they were given instead of by what they have made of themselves.

We now turn to genetic technologies that extend the prospect of behavioral prediction envisioned in *Gattaca*. The technologies include the simultaneous genotyping of upwards of five million genetic markers, and massively parallel DNA sequencing of the entire genome. Beyond what was imagined in *Gattaca*, epigenetics is a new frontier in which the genomic imprint of environmental experience can be measured to understand how combinations of genes and experience (gene by environment interactions) have affected a person. Are these new frontiers science fiction or science, and if they are science what are their realistic applications?

BEHAVIORAL PREDICTION, A SCIENCE IMPERFECT

Human behavior is notoriously unpredictable, yet psychiatrists and judges are routinely given the task of behavioral prediction, with life and death at stake. Should a potentially suicidal woman be committed to a psychiatric facility? Should we withhold bail for the man who hit his neighbor with a two-by-four? Should an unbalanced, and possibly schizophrenic young man who has posted threatening internet messages be forced to accept psychiatric assessment and treatment? Should he be allowed to purchase guns? Should a pedophile be given early parole given his exemplary behavior while incarcerated among adults? Should a woman receive a protective order against her former husband who in the past month twice showed up uninvited at her house? What should be the treatment/punishment for a drunk driver who has now been arrested three times in five years? Can an eight-year-old girl be safely returned to her parents? Is it appropriate to try some particular 17-year-old murderer as a juvenile?

DNA fingerprinting as used by forensic scientists identifies who we are, but the more subtle and potentially powerful side of genetic analysis is its ability to tell us what we are. How should the information that criminal behavior, addictive behaviors, and disease responses such as occupationally associated cancers and adverse drug reactions be used? For example, will whole genome epigenetic analysis, by which the impact of experience including stress and mutagenic exposures can be measured, one day transform the adjudication of cases in which plaintiffs allege adverse effects of exposure, or ask for leniency because life experiences have traumatized them and predisposed them to whatever crime they have been convicted of?

COMMERCIALIZATION OF GENETIC BEHAVIORAL PREDICTION

The day of genetic behavioral prediction has dawned, but while the beginning of that day looks beautiful and rosy-fingered to some, others see reason in the clouds on the horizon to take warning.

In commercial testing, words written speak for themselves. On the website "DocBlum – Nutraceuticals for the Millennium", Steve Allen, who hosted the TV program "The greatest minds of all time", vouched for Dr. Kenneth Blum as a man whose research on the genetics of alcoholism "would change the world as we see it today". A genetic marker outside and downstream from a dopamine receptor gene (*DRD2*) was proposed to cause "reward deficiency syndrome". This genetic marker and "nutraceuticals" to treat the syndrome were patented. The "clinically-proven, and all-natural anti-craving ingredient complex" is an "all-natural dopamine neuronal-release promoter" available as pill, powder, sublingual tablet, or in fruit juice, intravenously, intramuscularly or as an intrarectal paste. The genetic test and nutraceuticals, now marketed by Salugen, include CraniYums, "the world's first functional candy". "The companion genetic test analyzes genes to better understand the chemical imbalance." A disclaimer reads, "These statements have not been evaluated by the US Food and Drug Administration. Our products are not intended to diagnose, treat or prevent any disease."

Genetic prediction is touted for many purposes. Genetic testing company 23andMe generates "clinical reports" and "research reports" using DNA from a saliva sample analyzed for over half a million DNA markers. "Earwax type" might not appear at the top of one's list of useful pieces of information but without commenting on the strength of predictive validity of the reports, several others are relevant to the origins of behaviors and diseases. These include alcohol-induced flushing, which is protective against alcoholism but a risk factor in upper gastrointestinal tract cancer if one has it (as more than half a billion people do). Listed as "research reports" are alcohol dependence, heroin addiction, dyslexia, avoidance of errors, memory and pain sensitivity.

THE FUTURE OF GENETIC BEHAVIORAL PREDICTION

Seldom is DNA destiny and the predictive value of any individual genetic marker available today is low, but it is a misappreciation of the science to disregard the importance of inheritance. In the future, individual predictive genetic variants, some of which are already known, will be integrated with clinical history and neuropsychological testing to better define behavioral disorders, and better predict treatment response and outcome. They will also provide a more complete picture of the causation of human behavior. The goal of biological psychiatry, of which genetics is a part, is to improve treatment and prevention by identifying new treatments and targeting the right treatment to the right individual.

Overall, medicine is not so far advanced beyond its beginnings as we would like to think, and this is especially true for the common diseases responsible for most mortality and morbidity, and for aging itself. Physicians have thousands of specific cures to offer; however, there are many major problems and whole areas of medicine that, relatively speaking and like psychiatry, are still in the dark ages.

As discussed in greater detail elsewhere, the functional complexity of the brain is daunting, leading to ultimate limits to our understanding of the brain, and therefore its diseases. The brain may constitute the living proof of Goedel's theorem that a system is incapable of explaining itself; to do so would require the creation of a system even more complicated and thus more difficult to explain. As an aside, a variant of this dilemma occurs when someone tries to explain that God accounts for the origin of all things. The brain's complex development, plasticity and structural complexity are all factors that make it a very different and far more mysterious structure than any other organ in the body.

From the brain's complexity, it naturally follows that a genetic marker that predicts behavior will not necessarily be explanatory. In other words, it does not follow that linkage between a DNA marker and a psychiatric disease leads to an understanding of how the genetic variation alters behavior; or if the understanding of the intervening steps is achieved, comprehension may arrive years after the discovery and long after the genetic marker is used as a predictor. We may be able to predict whose brain works differently, but not be able to explain why. It's a little as if a policeman noticed that red cars are faster than blue cars. Perhaps the policeman begins to pay more attention to the red cars but what he hasn't figured out is that the drivers of those cars are a little crazier or the engines more powerful. Although the brain's complexity is beginning to be unraveled, it is likely that the mechanism of action of a genetic variant predicting brain function will remain mysterious long after we understand why some cars are faster than others.

As an example, and perhaps to alleviate any impression that might have been created earlier in this chapter that there are not genes whose roles in behavior have been well defined, we will accelerate into a discussion of real-world, clinically important gene discoveries. I will contrast what is known about a gene causing acute hemolytic anemia (specifically, G6PD deficiency) versus different genes causing aggression and self-mutilatory behavior (specifically, Lesch–Nyhan syndrome) and mental retardation (specifically, phenylketonuria, PKU). These three gene discoveries were made more than four decades ago, so there has been ample time to validate them in many ways and to work out an understanding of mechanisms. Also, these findings led to specific approaches to prevent or mitigate disease, and represent the sort of genetic information that physicians obtain and integrate into overall assessment and planning of care.

A GENE CAUSING ANEMIA

One cause of acute hemolytic anemia is inherited deficiency of glucose-6-phosphate dehydrogenase (G6PD). G6PD has several important functions, one of which is to regenerate NADPH, a molecule that can undo the oxidation of sulfhydryl groups found on many proteins of the cell. That is important, because otherwise the damaged proteins do not work. The red cell is particularly susceptible to oxidative damage that can occur after eating certain types of foods (e.g. the always-delicious fava bean) or taking certain medicines (e.g. antimalarial drugs). During the Korean War, a large fraction of African–American, male soldiers suffered massive destruction of their red cells when given chloroquin as malaria prophylaxis. It was noted that if the medicine was maintained, in time the hemolysis disappeared as a problem because the older and more fragile red cells had been eliminated. The soldiers then were able to maintain normal blood counts because the production of erythrocytes by the bone marrow had ramped up, a process that also came to be understood at the molecular level as caused by increased release of erythropoietin, providing the intellectual property basis for one of the world's first successful biotech companies (Amgen). Two very common genetic types of G6PD deficiency, one in West Africa and the other in peoples of the Mediterranean basin, were identified, and the mechanism for both was elucidated.

The mechanism of disease has been more elusive for genetic mental disorders even when the specific gene is known, and even when the molecular function of the gene is known. Two examples are Lesch–Nyhan syndrome and PKU. Similar to G6PD deficiency, both Lesch–Nyhan syndrome and PKU are biochemical genetic disorders for which we know the enzyme pathways. The genetic variants responsible have been exactly identified and we know how they impair or block the functions of the enzymes. Also, the effects of the enzyme deficiencies have been worked out at the molecular level and, for both, alterations in levels of the small molecules that the enzymes act upon have been precisely measured. The continuing mystery as to how the altered biochemistry translates into altered behavior reflects our relatively poor understanding of our complex brains.

A GENE CAUSING SELF-MUTILATION

Lesch–Nyhan syndrome is caused by deficiency of the HPRT enzyme. This enzyme is involved in the salvage of purines, which are important as building blocks for DNA and ATP, a molecule which plays a major role in cellular energy transfer. The disease was first recognized in 1964 by a medical student at Johns Hopkins, Michael Lesch, working with his mentor William Nyhan, who was a biochemical geneticist. When I was a clinical research fellow training at the National Institutes of Mental Health, I was fortunate enough to see several of these patients as part of a biochemical genetic study. Carl Merril and I were studying these boys precisely because they represented one of the very few instances at that time, almost 30 years ago, in which a specific genetic variation was known to cause a complex behavioral phenotype. When we arrived at the hospital where the affected boys were in chronic care, we saw that the sick boys were restrained in chairs. I knew that this was necessary because they tended to mutilate themselves, and would behave aggressively, especially when made anxious. They felt more secure – and this is neither condescension nor rationalization – when restrained during an anxiety-provoking situation such as our visit. We had not touched them, and they were emotionally calm and not unhappy, but nevertheless they grimaced and their arms writhed in their bonds. Sadly, the boys had severely mutilated the ends of their fingers. When I drew the blood sample from the forearm, one of the boys pushed his arm towards the needle.

The behavior of these boys was just one consequence of a biochemistry that involves the unleashing of the synthesis of purines from normal restraints. Although Lesch–Nyhan syndrome is rare, it shares biochemistry and non-behavioral aspects with the much more common disease of gout. Gout was known to the ancient Egyptians and recognized throughout medical antiquity as the arthritis of the rich and powerful. The mechanisms of gout are understood. Purines key to both gout and Lesch–Nyhan syndrome are synthesized in the body but also derived from the diet, particularly foods rich in DNA, and especially dark muscle, fish and other meat that is rich in mitochondria, which have their own DNA genome. An excess of intake of these foods leads to an excess of a molecule called uric acid, a metabolic product of purines. Uric acid is normally excreted by the kidneys, and to a lesser extent the gut, but high concentrations cause a problem, whether due to diet or genetics. When the pH drops (acidity increases), crystals of sodium urate precipitate from solution in the joints (especially the big toe), kidneys, bladder and other tissues. Visible crystalline deposits in the flesh are called tophi. As one might imagine, joint pain from the needle-like crystals and accompanying inflammation is severe. The cause was discovered by Antonj von Leeuwenhoek, who was born Thonis Philipzoon in 1632 but who renamed himself "Anthony from the Lion's Corner", thus illustrating that you have to be pretty careful when choosing a new name for yourself. Illustrating the power of technology (an MRI scanner is said to be worth a roomful of neurologists), von Leeuwenhoek applied the microscope, a recent invention of Hook's, to a series of problems. He made a series of amazing discoveries: protozoans, red blood cells, the compound eyes of insects, microparasites, sperm, parthenogenesis, bacteria in dental tartar, blood capillaries. As the discoverer of a whole new microscopic world, he became world renowned despite never having published a scientific paper or presented his work at meetings of distinguished scientists. In 1679 he observed the distinctive needle-like crystals of sodium urate in tophi from a gout patient, and correctly guessed that these were the cause of the pain.

However, as compared to gout, Lesch–Nyhan syndrome is more severe and includes behavioral and cognitive consequences. Although *HPRT* was the first gene tied to behavior at the molecular level, we still do not know why the biochemical problem leads to the behavioral problem. The gene is on the X-chromosome and the rare variants act recessively. Therefore, severe HPRT enzyme deficiency occurs in males, who carry only one copy of the gene, unlike women who carry two copies. The predicted frequency of such an X-linked recessive disease in females is the square of the risk of the disease in males – about one in a million times one in a million. The purine salvage pathway enables the body to recycle purines, and through a negative feedback process *de novo* purine synthesis is inhibited. If purine salvage is blocked by HPRT deficiency there is a large increase in the *de novo* synthesis of purines, and there is an increase in the metabolites of purines and changes in levels of adenosine and other molecules that can alter brain function. Infants with the disease can be spotted because of crystals of uric acid in the diapers. They endure severe gout and renal stones, moderate mental retardation, poor muscle control, spasticity and hyperreflexia. There is involuntary grimacing and writhing movements known as choreoathetosis. Most never walk. Drugs that treat the hyperuricemia, such as allopurinol, address the problems of gout and kidney stones. However, at two to three years the behavioral problems of self-mutilation and aggression emerge, beginning with biting of lips and tongue, and headbanging. These behavioral problems are increased by stress and found in about 85 percent of patients, such that most have their teeth extracted. Furthermore, nearly all require restraint.

A GENE CAUSING MENTAL RETARDATION

PKU is a less dramatic behavioral genetic disease, but more abundant and equally mysterious. PKU is usually caused by deficiency of phenylalanine hydroxylase, which metabolizes the amino acid phenylalanine. It is also more rarely caused by deficiencies of enzymes that synthesize the main cofactor that this enzyme needs to function. If one has PKU, high levels of phenylalanine and the metabolite phenylpyruvic acid build up in the body, and most importantly, a metabolite gets into the brain. The eyes tend to be blue, the hair blond, and if the disease is unchecked, there is the development of cognitive deficits. However, those deficits can be prevented by limiting phenylalanine in the diet, a course made possible by detection of PKU by the early screening of infants, as will be described. PKU is a common, diagnosable and treatable genetic behavioral disorder.

WHAT IS A GENETIC TEST?

When I was a young scientist interested in understanding genetic metabolic diseases affecting the brain I read a letter in the Washington Post by the late Seymour Kaufman, an authority on neurotransmitter biochemistry. Seymour happened to be a section head in the laboratory led by Giulio Cantoni (the discoverer of the methyl donor *S*-adenosyl methionine). I disagreed with what Seymour had written but naïvely hoped that he would be glad to know that someone had read his letter, and cared (in science all publicity is good – unless you are an NIH scientist who has not reported an outside activity). I was amazed that his opinion was that PKU was *not* an example of genetics unraveling a behavioral disease. For his part, Seymour was unimpressed by what I had to say and courteously stuck to his guns. Our pleasant disagreement ever since influenced me to think about the nature of genetic behavioral disease.

Kaufman's argument was that the origin of PKU was discovered via biochemistry and enzymology – studies of small molecules and proteins instead of DNA. I was sure that this was a false distinction; although PKU was not a DNA-based discovery it was a prime example of biochemical genetics in action. Perhaps someone who reads this book will draw the opposite conclusion, forming another of those scientific circles of contradiction. If so, good! However, ever since reading Kaufman's letter I persisted in my thinking that the transmission of an enzyme deficit is genetic in the same way that transmission of a blood marker such as ABO detected via an antibody reaction is genetic, even though neither is DNA based. After all, to work out the fundamental laws of inheritance, Mendel used visible characteristics of peas, and knew nothing about the DNA or the proteins and small molecules that are closer to the level of DNA action than the visible genetic characteristics he relied upon.

Misunderstanding about what is and is not a genetic test continues to be widespread. Genetic testing and tests with genetic implications are frequently unrecognized as such. However, addition of a DNA component to the test would immediately lead to concerns about genetic testing even if it adds no new information. For example, if a man is diagnosed with hemolytic anemia due to G6PD deficiency (an X-linked disease usually transmitted from unaffected carrier mothers to affected sons and unaffected carrier daughters) then it is almost certain that his mother was an unaffected carrier. The diagnosis of G6PD deficiency in the man may have been made clinically or via measurement of his enzyme level but there is a powerful genetic implication for his mother, and any of his male siblings could immediately be known to have at least a 50 percent risk of G6PD deficiency (the 50 percent chance of inheriting the mother's X chromosome carrying the disease allele), with no testing performed at all. This is an example of a functional genetic test, with all the usual implications including ethnic genetics, G6PD deficiency being far more common in Africans and southern Europeans than in northern Europeans or Asians. There can be no perfect boundary between the disciplines of genetics and cellular molecular biochemistry: genetics illuminates biochemistry and understanding the molecular pathways and networks of the cell is the key to the real endgame of genetics: the isolation of the functional locus and its mechanism of action. However, in the case of PKU it was a genetic disease that was being unraveled, by the means available.

Although measurement of phenylalanine hydroxylase enzyme activity is itself a genetic measurement, the screening for PKU in infants has been accomplished even more cleverly and efficiently for the past half century by taking advantage of a genetic glitch in the genome of another life form, via the *Guthrie test* (although this test is now being superseded by other methods). In bacteria, a mutation was identified that eliminates their ability to synthesize phenylalanine, which is required for synthesis of the proteins they need. Bacterial growth therefore depends on external phenylalanine supplied by blood spots. In the Guthrie test, the growth of the bacteria is measured in the blood of newborns spotted onto filter paper. The blood of babies with phenylalanine hydroxylase deficiency, and related mutations, has high levels of phenylalanine, enabling the bacteria to grow into a larger diameter colony. This test is not perfect (only 10 percent of infants who test positive actually have PKU) but is extremely rapid and inexpensive for screening, and afterwards can be followed up with more detailed studies. A small fraction of infants with PKU have defects in enzymes synthesizing biopterin, a cofactor for the phenylalanine hydroxylase enzyme. PKU screening, and the follow-up dietary restrictions, prevent more than 27,000 cases of mental retardation annually, in the USA alone.

Summarizing, can specific genes predict behavior? Yes. Next, does that yet have profound implications in medicine, in the courts or in the lives of people who may want to know more about themselves? I think the answers are probably yes, with limitations. Are we ready to target "nutraceuticals" or pharmaceuticals to specific individuals based on their genotypes? No. We will see that genetics of behavior is in its infancy, with few functional loci identified that can predict behavior and only a small portion of the overall variance in behavior, from person to person, accounted for. However, there are also several functional genotypes that themselves are not highly predictive but are more strongly predictive in the context of an environmental exposure, usually stress.

13

Warriors and Worriers

"An eye for an eye makes the whole world blind."

Mahatma Gandhi

Why should there be common functional genetic variants that cause some people to behave differently from others? The reason is that societies composed of varieties of people are more successful, but at the level of the individual selfish gene the explanation is that the gene for a behavior may be selectively advantageous when the behavior is rare, but disadvantageous when the behavior is common, and that particular variants influencing behavior are advantageous at certain times and in certain niches, but not in others. In this chapter we will discuss this mechanism: balanced selection. It has maintained two common forms of a gene that influences both cognition and emotion, and in these two behavioral domains the advantages of the two forms of the gene counterbalance, leading to "warriors" and "worriers", each of whom may find their niche within society. Evolutionarily, the advantage goes to the individual who is best suited to occupy an available opportunity in the community, but the result is that no human society is entirely composed of Attilas or Gandhis. We are both warriors and worriers, and fortunately the balance has sometimes worked, after seasons of violence and vengeance.

Our Genes, Our Choices
DOI: 10.1016/B978-0-12-396952-1.00013-0

A COMMON GENE "FOR" COGNITIVE FUNCTION

To understand why there are genes that influence behavior, it has to be understood that many common genetic variants can be surprisingly easily tied to behavior because these gene variations exist for the purpose of causing behavioral variation, and not just to cause variation in an obscure biochemical indicator. In other words, we can observe the behavioral effect and not just the effect on a biochemical pathway not because we are clever but because the genetic variant exists in order to cause behavioral variation of the type we observe in everyday life. *Evolution acts through biochemistry to alter behavior, but does not select variants whose only effect is to alter biochemistry.* The variant may exert a stronger effect on the biochemical intermediate phenotype, but it exists to alter behavior.

There is a common genetic variation in catechol-O-methyltransferase (COMT), an enzyme that influences both cognition and emotional resilience. It serves as a beautiful illustration of the way that evolution can create common genetic variation, such that as a result of a difference in a single gene one person differs from another in multiple behavioral domains. The polymorphism involves two alternative amino acids at position 158 in the protein sequence: Val158 or Met158. The effects of the alleles counterbalance each other and as a result both alleles (genetic variants) are maintained at high frequencies (40–60 percent) in populations worldwide.

COMT was discovered by Nobel laureate Julius Axelrod – a genius of neurochemistry who began as a technician but ended up training or inspiring a generation of neuroscientists, including me. One exciting thing about training at the National Institutes of Health (NIH) is that as a young scientist I ran into Julius Axelrod a few times, and he was invariably kind. However, I also came close to running him over as he jaywalked in a preoccupied fashion. Thank goodness I didn't hit Axelrod because I doubt anyone would have believed, "Julius Axelrod darted out in front of my car".

In the brain, the main function of COMT is to metabolize catecholamine neurotransmitters. COMT provides one mechanism by which the action of several neurotransmitter molecules is ended. In the frontal cortex, the action of dopamine is particularly dependent on COMT activity because another mechanism, the dopamine transporter, is not active there. The Met allele is less effective than the Val allele, leading to higher levels of dopamine in the brain's frontal cortex, especially in people like me (about one in six of the population), who have two copies of the Met allele (Met158/Met158 homozygotes). Several years ago, a team that included me, members of my lab and Daniel Weinberger's lab at the National Institute of Mental Health found that people with this Met/Met genotype (all other things being equal!) tend to have better frontal cortical function than people with other genotypes. This translates into better performance on cognitive tasks that are executed by the frontal lobe.

EXECUTIVE COGNITIVE FUNCTION

In combination with genotype, performance on any of these tasks might be used to predict behavior influenced by frontal lobe function, and so I will briefly describe the tests. Those tasks include working memory, which is the ability to remember items for short intervals of time. Working memory can be measured in a variety of ways, for example by having a person repeat digits forward or backward. Good performance on the digits reversed task might be seven digits – not easy! Another measure of working memory is the N-back task, in which the person is asked to recall a stimulus presented one event ago (1-back), two events ago (2-back), etc. As the task becomes more difficult, the metabolic activity of the frontal cortex intensifies, stress increases and performance eventually declines. When the task is easy (for example the 1-back test) and everyone is performing the task accurately, a person with the Met/Met genotype is still different in an important way – showing lower metabolic activity, reflecting their higher *cortical efficiency*.

The frontal cortex is also the master switchboard for task management – enabling people to apply the cognitive strategy appropriate to the situation – and it is thus the executive cognitive center of the brain. This latter ability is crucial to the ability of people to make adaptive responses to their environment, so that if one approach is not working they can switch to a different strategy instead of making perseverative errors. There are many measures of executive cognitive function. Two that are often used are the Stroop Test and the Wisconsin Card Sort Test. The Stroop Test is often illustrated by speakers at scientific meetings, and when that is about to happen I try to prepare my friends. Typically the speaker flashes a slide such as the one in Figure 13.1 and asks the audience to call out what it sees. Most of the audience is silent. Some dutifully call out, "Red". A few oddballs call out, "Blue". But someone wearily says, "Stroop test", having classified the stimulus as a slide illustrating this frontal cognitive task. The Stroop test relies on the fact that "Red" can be classified as either a word or a color

FIGURE 13.1   What do you see? A test of frontal cognitive function.

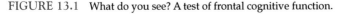

(and several other things besides). In the actual neuropsychological test the interference between cognitive processing as word versus color is measured using response time, the word usually more greatly interfering with ability to name color and increasing the response time when there is mismatch. As just mentioned, someone's brain might also process the stimulus as a cognitive test, or in a myriad of other ways, many of as little specificity as "color" (photons, word, slide, and so on). Regardless of how well one task-switches, or how strangely – there is a term in psychiatry called "loose association" – it is the frontal lobe that enables us to task-switch from one classification strategy to another. Also, it is obviously not a good overall cognitive strategy to use too many cognitive strategies – this could be a formula for slow decision making or cognitive paralysis of the type that can trouble a human brain with its potential for imagining many meanings.

COGNITIVE FLEXIBILITY

Perhaps some of you are by now getting the idea that scientists who measure cognitive flexibility are not always very cognitively flexible, but probably that conclusion is unfair. To access the way people think, they use measures that tap into how people usually think. To produce objective and reliable data on cognition across many different types of populations, we require simple, standardized tests. We probably have little understanding of what algorithms a person is running in order to solve problems but we can at least determine what parts of the brain are activated during cognitive tasks, thus isolating the locations of the circuits involved. The amount of time required for the decision making and the sequence of involvement of brain regions tell us more. We are able to discover genes, neurotransmitters and drugs that modulate these processes. We can define that certain individuals are better at performing a task than others and that two people who perform equally may be working at very different levels of intensity. This is seen, for example, when brain frontal metabolic activity is measured by functional magnetic resonance imaging (fMRI) during tasks that depend on frontal lobe function.

Some people's frontal cortices are "poorly tuned" such that they require more metabolic effort, reflecting more neural activity merely to perform at the same level. Frequently, schizophrenic patients and their well siblings show such frontal cortical inefficiency, and it can be manipulated pharmacologically, for example by augmenting dopamine. But we actually have no idea what it means to be frontal cortical inefficient. What is the brain doing that it appears to be inefficient? I would like to believe (and this is an admission that what I am saying next is speculative) that sometimes cortical "inefficiency" is a manifestation of alternative patterns of cognition – including more rigid and more flexible patterns – and that such alternative patterns may be invaluable under different circumstances than represented by a particular test, or perhaps any test devised.

The difficulty of assessing cognitive flexibility is well illustrated by the inflexibility of some of the tasks that have been designed to test it. For example, let's reconsider the Red version of the Stroop test. This is a fine illustration of the greater range of potential responses that are possible. For example, is there anything *really* wrong with the response: "one-syllable word"? It is a generic answer, but so is the acceptable answer "blue". The answer is that these responses may be a little weird or unusual, but they are intrinsically accurate, and in fact could be called for in certain situations. Failure to understand the true range of potential accurate responses is one way that we underestimate human intelligence, and underestimate the decision making that our brains constantly perform.

We covet cognitive flexibility, but do not do well at encouraging it or assessing it. Starting early in school, teachers may not only look for the "correct" answer, but at times punish a child who arrives at the answer by a different method. This was a cardinal error of "new math". Or, we want to assess the cognitive flexibility of a job applicant. Can he "think outside the box"? A typical question – the answer to which is found on websites – is:

Why are manholes round?

One applicant sits silently, even though he knows that "playing possum" is not a good strategy in a job interview. Another compliments the interviewer on his question, and indicates enthusiasm for the general concept of thinking out of the box. This could be a good idea when interviewing for an "assistant" position. Next is the person who gives the expected, out-of-the-box answer, "Because round manhole covers don't fall into manholes when turned this way or that as compared to square covers" (as if a square was the only alternative shape). So all well and good. This applicant is "creative". However, the next interviewee answers, "Because round covers don't fall into manholes when turned this way or that and land with a pointy end on the top of someone's head". Too much information, but nevertheless everything may be okay, as long as no competitor responded exactly as desired. However, what is to be done with the applicant who has too many explanations? She speculates that round covers are easily rolled into place, suggests that viewed from above human bodies are roughly cylindrical and the manholes we fit them into are generally round (the round peg/round hole theory), leading to the fact that round covers are cheaper to produce because they conserve iron and being lighter are cheaper to ship. Or perhaps it is because round covers fit whichever way one orients them, thus making it simpler to replace them. Perhaps it is a tradition. And so on.

PERSEVERATION

Einstein observed that the definition of insanity is doing something that does not work over and over and expecting a different result. However, people frequently perseverate and psychologists have ways of measuring the tendency to perseverate, and perseveration errors. The Wisconsin Card Sort Test measures executive cognitive flexibility in a different way than the Stroop Test. The subject is asked to match one of several test cards to a target. For example, the target might depict a halibut (a type of fish) and the test cards might include a cat, a fisherman and a whaling ship. The person might match the cat and the fisherman to the halibut because all are animals. Or the person could match "fishing cards" to the halibut. Whichever matching strategy is used the person is informed that the answer is wrong. People with frontal lobe damage perseverate, continuing to use the strategy that has been labeled incorrect on the next test in the series. In a remarkable finding that has been replicated by many others and in many types of populations: "normal controls", patients with schizophrenia, siblings of schizophrenic patients and head-injured patients, we found that people with the less efficient Val158/Val158 genotype tended to make more perseverative "errors". It is particularly interesting that patients with schizophrenia and head-injured patients show the genotype effect on cognition, because both of these clinical populations already tend to have deficits in frontal lobe function, as manifested by their ability to perform the Wisconsin Card Sort Test.

Around this time one of my sons, Ariel, and I traveled in the Middle East with Danny Weinberger, Danny's son Collin and his wife Leslie. Danny is perhaps the world's premier "biological psychiatrist" and an expert on schizophrenia, and I had joined with him to make the first discoveries using "imaging genetics". One key to schizophrenia is the function of the frontal lobe of the brain, and a major determinant of frontal lobe function is the neurotransmitter dopamine. Frequently, the function of the frontal lobe can be augmented by boosting frontal dopamine levels, as happens if one takes the drugs amphetamine or methylphenidate. Thus it is that methylphenidate is used in the treatment of children with attention deficit hyperactivity disorder (ADHD), and it often helps.

While in Petra, a Jordanian taxicab driver gave Danny and me an interesting illustration on how dopamine works in the frontal lobe, helping to explain why a deficit in the activity of the COMT enzyme would augment dopamine levels and frontal function. It turns out that unlike most neurotransmitters dopamine diffuses far from the particular synapse where it was released. This is especially true in the frontal cortex, which largely lacks a dopamine transporter that takes up released dopamine so that the neurotransmitter can no longer bind neuronal

receptors on either side of the synapse. So one may say that the aim of
the cell releasing the dopamine does not have to be very good. It will
hit a target even if released anywhere in the general vicinity. I noticed
a large handgun on the front seat of the taxicab. Much to Danny's chagrin I asked the driver about his "conversation piece". It turned out that
he was a "retired" intelligence officer. Then I remarked that this type
of handgun was not very accurate. There was a moment of silence. The
driver stopped the cab, waved the gun at us and said, "At this range
it's accurate". Later in the trip, and after run-ins with Sami the Guide,
who destroyed my camera, camel races, a scuba patrol led by a com-
mando and a make-believe encounter with the Dwarf Rabbi of Tel Aviv,
we found ourselves back at the border hoping to reenter Israel. The cute
Israeli officer suspiciously eyed my diving knife and with a meaningful
look asked Danny, "Are you traveling with each other?" "We are traveling *against* each other", I said, and from then on that's how Danny
described the trip, which to be serious was one of the best times of our
lives, and definitely an out-of-the-box experience.

Following our travels together in the Middle East, Danny and I
did not seem to work together as often as before. His NIMH lab went
on to make a very important additional discovery about the effect of
the COMT polymorphism on cognition, which was that the cognitive
advantage of the Met158/Met158 genotype disappeared under some
circumstances that often occur in everyday life. The reason is shown in
Figure 13.2. At a particular dopamine concentration frontal cognitive
performance is maximal. Under ordinary conditions, a person with the
Met158/Met158 genotype tends to have higher dopamine levels and

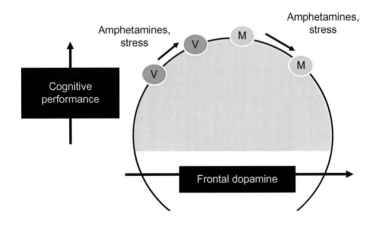

FIGURE 13.2 COMT variants (Val 158 and Met 158) and the inverted U effect of
dopamine on frontal cognition.

closer to this cognitive optimum. However, the relationship between dopamine concentrations and cognitive performance is an inverted U-shaped curve. Too much of a good thing is a bad thing. When they take amphetamine-like drugs that boost dopamine levels and improve cognition in Val158 carriers, the dopamine levels of Met158/Met158 homozygotes become excessive, and their performance deteriorates. Stress, which elevates levels of dopamine and other neurotransmitters, can lead to the same thing.

WARRIORS AND WORRIERS

The whole world will not, as Gandhi feared, go blind, because some of us are by nature worriers, and some are warriors. As we have just seen, Met158/Met158 homozygotes turned out to have a cognitive advantage, but a disadvantage in cognitive resiliency under stress. At about the same time as Danny and I, and others in our labs, were discovering the advantage of the Met158 allele, I and other colleagues were discovering the major downside to the Met158/Met158 genotype. The finding was that individuals tend to be less resilient in the spheres of emotion and pain sensitivity. The lower pain threshold of Met158/Met158 homozygotes was more dramatic, and was observed by measuring people's pain threshold, but was most dramatic using brain imaging to measure the brain's response to pain. Jon-Kar Zubieta at the University of Michigan had shown that pain threshold is predicted by the brain's ability to release endorphins (the body's natural, endogenous opioids) after a painful stimulus. As shown in Figure 13.3, opioid receptors are found throughout the brain, especially in regions of the brain that are important in the pain pathway, such as the thalamus, but also in the limbic system. The limbic system, consisting of a series of regions including the amygdala, hippocampus and cingulate cortex, regulates emotion. Emotional regulation is crucial in the interpretation of pain: people in emotional distress have lower pain thresholds. Also, chronic pain directly leads to emotional problems, especially depression and anxiety.

To measure pain threshold, hypertonic saline (salty water) was infused into the jaw muscle via a needle. This results in a burning pain which can be carefully calibrated so that the pain threshold can be accurately measured. People with the Met158/Met158 genotype had lower pain thresholds and stronger emotional responses to the equivalent amount of pain. They were very poor at releasing endogenous opioids following the painful stimulus. Probably their endogenous opioid release was already maximal just from being placed in the experimental situation. These then are the "worriers", cognitively advantaged but less resilient to pain and emotional distress. In contrast are the more

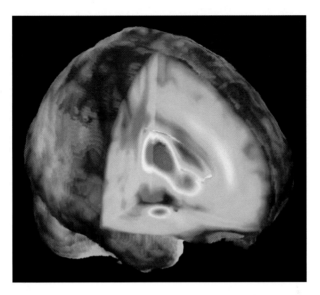

FIGURE 13.3 Opioid receptors are found in brain regions involved in pain and emotion. *(Mu opioid receptors imaged by [¹¹C]carfentanil positron emission tomography) (Image courtesy of Von-kar Zubieta)*

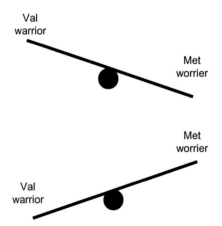

FIGURE 13.4 Top: low stress – advantage goes to Met158/Met158 homozygous individuals; bottom: high stress – advantage goes to Val158/Val158 homozygous individuals.

resilient Val158/Val158 "warriors" (see Figure 13.4). Probably from the earliest origins of humans both types of individuals have been needed, and selection for these genotypes was frequency dependent and also depended on just what pressures the social group was experiencing. If a tribe lacked either warriors or worriers it was potentially in trouble, and because ancestral social groups were small it was necessary that the genetic variants producing the alternative behavioral patterns would be relatively common. One hunted and another tilled the soil and kept the fire, and both needed to be good at what they did. Therefore, it is almost certainly incorrect to view common functional genetic variants affecting brain function as having been selected for the purpose of creating disease; however, it is probably no accident that they alter behaviors that are readily observable: they evolved for exactly that purpose, and are genes "for" behavior.

14

How Many Genes Does it Take to Make a Behavior?

"It takes all kinds to make a world."

Don Quixote, **Volume 2, Chapter VI, Miguel de Cervantes (loosely translated)**

Do genes that alter complex behavior act alone or in combination? Can one gene variant, like Sauron's ring that ruled over the rings of man, dwarf and elf, overwhelm the actions of other genes? More specifically, is one genetic predictor sufficient, or will they have to be used in combination? As we will see, it depends. Epistatic means due to the combined action of genes, and if there are many interacting genes the trait is often said to be polygenic ("many genes", although there is actually a distinction between the number of genes and whether they interact additively or epistatically). Oligogenic means due to the action of a few genes and monogenic means one. If you asked most of the geneticists I know, they would most likely tell you that any complex behavioral trait, including impulsivity, would be epistatic and polygenic. Also, almost any neurobiologist would tell you that this is how behavior has to work. However, they are wrong.

EPISTATIC AND POLYGENIC MODELS OF BEHAVIOR: ON EPISTASIS

In the polygenic model, and here I am specifically speaking of non-additive (or epistatic) effects of genes, one genetic variant is insufficient to alter behavior; it always takes a combination of several variants (which usually are conveniently unknown!). Thus, the polygenic model resembles the method used by the Joker, Batman's nemesis. The Joker terrorized the citizens of Gotham City by having Axis Chemicals manufacture cosmetics that were individually harmless but fatal when combined, thus producing "Gotham's shopping nightmare". Batman used a powerful computer located in the Bat Cave to figure out which combinations of toxins were causing the problem. In genetics, this type of gene–gene interdependence of action is called *epistasis*, and it is illustrated in Figure 14.1, using puzzle pieces to represent the functional variants that can act together in different combinations. People who have four or more of the puzzle pieces usually have the complex behavior. People who have only one or two usually do not.

For a geneticist, one of the most annoying things about people who are highly knowledgeable about behavior and its neurobiological basis is that they are usually certain that the complexity of the brain will prevent any individual genetic variant from exerting a significant effect on behavior. If true, the consequence of this is that genetic prediction is greatly complexified: if one has discovered the genes they will only be useful if analyzed as combinations. Also, it is obviously a lot more difficult to find the genes in the first place.

Margaret McCarthy, now Professor in Physiology and Associate Dean at the University of Maryland, used to patiently explain that the brain's

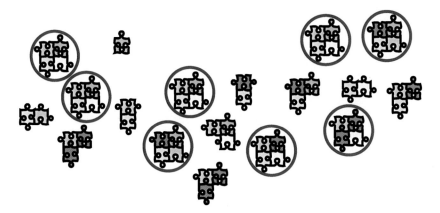

FIGURE 14.1   Polygenicity (epistasis).

systems can compensate for excesses and deficiencies. Even if a gene would influence a neural system to make a person more anxious, a second or third neural system could moderate the behavior, with the net result that there would be no effect of a single genetic variant. From this perspective, claims for the effects of single genes would be "startling" and require stronger proof than findings that really only are telling us something that we already know. Thus a general skepticism about single gene findings in behavior could essentially cause us to reject almost all such findings. Is that fair? Unfortunately, the answer is "Yes!"

BAYESIAN REASONING: HOW TO USE PRIOR PROBABILITY

Some 250 years ago, mathematician and Presbyterian minister Thomas Bayes formally proposed a method of truth-seeking that people have always relied upon in everyday life: a posterior likelihood depends not only on an observation but on the prior likelihood of the thing. It is not only okay to doubt the absurd. It is an obligation. A cat is pulled out of a hat, but nevertheless and despite the evidence of the senses one can continue to doubt that cats live in hats or are so well trained as stay put when carried about by a magician. Here though, I should report that while visiting St. Louis for a medical residency interview I did smuggle three cats into the historic Saint Louis Renaissance Grand Hotel, all cozily ensconced in a spacious, carefully handled suitcase. Probably the porter knew that it was impossible there could have been cats in that suitcase. At the magic show, we assign a low prior probability to the proposition that a cat is in a hat. The observation of a cat being pulled from the hat, and failure to detect the sleight of hand, constituted some positive evidence. However, the evidence was probably insufficient to alter your view that cats do not live in hats. The Bayesian calculation of the posterior probability, taking into account the old information and the magician's trick, would be as follows:

> Prior probability (before the magician's trick) of cat in hat: low, for example 1 in a billion (note: it's a small probability – and so small that one does not need to estimate it precisely, nor is it easy to estimate precisely).
> Evidence from magician's trick that cats live in hats on some regular basis: 1000:1.
> Posterior probability: 1 in a billion $\times$ 1000 = 1 in a thousand. In other words, we leave the show thinking that it is more likely that the magician has tricked us than that cats live in hats.

Suppose we watched the magician pull off the same trick ten times in a row. Some people would then actually change their opinion about hats and cats, and if you are one them please send name and bank access codes to 15 Rupees Plaza, Silver Springs, Maryland, which is where I instruct Nigerian email scammers to send their "supplicatory gift". We should retain our skepticism, realizing that the subsequent demonstrations do not represent independent "observations". If there was deception (the tricky magician problem) or something otherwise amiss with the first cat sighting, then all that followed could be subject to the same systematic error. The wise person also would not be more convinced if ten independent magicians produced cats from hats because we recognize that the magicians use the same tricks, even if we have not researched how they perform them.

Unlike the theater of magic, in the arena of science tricky behavior is fraud or misconduct. However, it is important to point out that truly fraudulent behavior blends gradually into lesser degrees of data fudging, distortion, selective reporting and cat-smuggling, which is one reason why some scientists are good at "pulling cats from hats". Of course, one cannot judge by success: some scientists are better than others and some are luckier: there are discoveries to be made and someone has to make them. Often it is the scientist who is led astray by a combination of wishful thinking, systematic error and "winner's curse". For example, even if the magicians were completely sincere about believing that cats lived in hats, and not in the least tricky, they still might all be using the same defective method. Some consistent aspect of their procedure might have allowed wily felines to get into their hats just before the crucial moment, or it might have happened just that once ("winner's curse"). In science we realize that multiple results achieved by exactly the same method may all be in error. It is a scientist's compensating ability to design experiments that rule out systematic error and chance findings through methodologically independent cross-validation that separates good science from science that is mediocre or bad. We cannot always get independent validation but we always want it. If the observation was true, then cats should be found in hats collected in other settings (this is called *predictive validity*). An investigator could be dispatched to hat-shops, and private collections of hats could be surreptitiously raided for analysis. Consignments of hats of all types could be collected from the four corners of the world. The hats would be shredded, processed and analyzed for any evidence that cats were present (cat hair, cat DNA, cat toys, etc). Also, the scientist making the case for cats in hats could provide evidence for the *mechanisms* by which they live there: what they eat, how they breed, and so on. This type of step-by-step mechanistic understanding, with observations connecting the whole, is the way theory is turned into fact. Alternatively, Bayesian skepticism might have led many of us to use our time more constructively.

So now we return to a more careful evaluation of the single gene model in behavior. Can it hold up against the *epistatic* model that appears to follow so naturally from the complexity of the brain and the dependency of brain development and function on the expression of more than 10,000 genes? Is the single gene model of behavior a "cat in the hat"?

BEHAVIOR AND THE SINGLE GENE

First, is the prior hypothesis that a single gene can alter brain function equally strong as the polygenic hypothesis? The answer is "Yes". The effect of a single genetic variation on brain function is usually nil, but the same goes for most combinations of variants; for one thing, the more extreme multigene combinations occur too rarely to account for common psychiatric diseases. For example, the likelihood that a person would have all of eight variants, each with a population frequency of 5 percent, would be 1 in 10^8, about one in one hundred million: $(2pq)^8$, where $p = 0.05$ and $q = 0.95$).

Also, there is every reason to expect that brain function can be altered by a single variant. First there is the evolutionary perspective offered for the *COMT* Val158Met polymorphism: this common polymorphism is maintained because it does alter behavior and if it does so on its own it will be better maintained by selection than would some variant that is required to be in epistatic interaction with others. Furthermore, the Bayesian perspective that there is a high probability that an individual glitch can disable a brain is supported by examples of such genes that have already been identified, and by examples of how glitches can alter or disable complex systems. Complexity can shield a system from harm but can also expose it. For a space shuttle, one missing heat shield tile can lead to the deaths of a whole crew of astronauts. In our bodies, single genetic variations lead to numerous monogenic diseases such as albinism, in which body pigment is totally lacking, or achondroplastic dwarfism. The characteristics (phenotypes) that the monk Gregor Mendel selected for his seminal studies that established the laws of inheritance were perhaps carefully selected, because they are all monogenic traits.

However, the observation that single genes *can* cause complex traits does not mean that monogenic traits are caused by a single gene: different genes and different genetic variants at the same gene may cause the trait in different individuals. The way that not one but many different single gene variants can be individually sufficient to cause a complex behavior is shown below. This model of gene action is called *genetic heterogeneity* (Figure 14.2).

This genetic heterogeneity does take into account the complexity of the brain. Since there are so many genes that influence a behavior, the genetic glitch that causes the behavior is very likely to differ from one person to the next. Even in the same gene there could be different types of glitches, some more severe and some less severe.

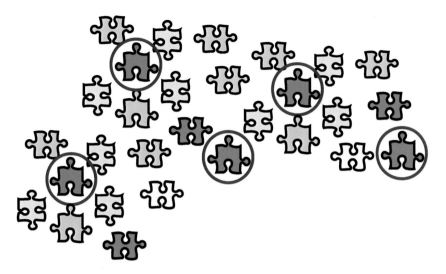

FIGURE 14.2   Genetic heterogeneity.

The truth as to whether behavior, and what proportion and types of it, is influenced by genes acting epistatically or by single gene effects will be determined by discovery of the genes. It is also being influenced by our increasing ability to measure phenotypes that are closer to the action of single genes, and that do not represent analgams of traits. The genetic transmission of many behaviors and psychiatric diseases in families is most compatible with additivity and single gene effects, although the "jury is still out". However, the answer falls within the realm of empirical science rather than theoretical or philosophical inquiry. Meanwhile, several chapters of this book have been devoted to discussion of single genes that do individually influence human behavior. In other words, genes *for* behavior. Because of the discovery of the monoamine oxidase A (*MAOA*) and 5-hydroxytryptamine receptor 2B (*HTR2B*) stop codons as causes of impulsivity, and the influence of *COMT* on resilience and cognition, we know that a single gene *can* be sufficient. It is also likely in other instances that dyscontrol behavior is polygenic. For example, someone with the less deterministic low-expression *MAOA* short tandem repeat (STR) allele might also have a second or third genetic vulnerability allele.

The Genesis and Genetics of Sexual Behavior

"No matter how much cats fight, there always seem to be plenty of kittens."

Abraham Lincoln

Our Genes, Our Choices
DOI: 10.1016/B978-0-12-396952-1.00015-4

Sex and mating are among the most varied and tightly regulated behaviors in the animal kingdom, yet they are surprisingly subject to variation. In humans both sex (gonadal sex) and gender (sexual behavior) are products of neurogenetic determinism as well as experience, but as will be seen within many species there is great diversity and flexibility in sexual adaptive responses. Much of the behavior of humans and other species of animals orbits around sex. We are sexual creatures. As contrasted with analysis of the origins of behavior of a cybernetic entity such as the HAL 9000 computer, which presumably has other issues, it is necessary to discuss sex in order to thoughtfully discuss human behavior and its genesis. In discussing instances from elsewhere in the animal kingdom in which genes determine sexual behaviors, we can learn how gene and environment influence sexual behavior. Sexual behavior and the effects of genes vary from species to species and are not being equated with the human condition. Sexual behavior in humans is accompanied by romantic feelings, fantasy, lifestyle differences and other nuances that are unrepresented in other species. Other species have unique aspects of sexuality that humans do not enjoy. Sex is a serious subject and the topic of much scholarly research. In discussing sexual orientation I use the homosexual/heterosexual terminology instead of gay/straight. "Gay" is shorter, better for wordplay not intended here, and is in common usage. However, "gay" is an unscientific term. It contrasts poorly with "heterosexual", for which a good substitute ("straight?", "non-gay?") is unavailable. Finally, we are still locked in cultural wars about sexuality, but sexuality is a single descriptor of a person. People are more multifaceted than that. If anyone is upset at the tone, nomenclature or content of this discussion of sex, I offer an advance apology. Any perceived insensitivity is unintended, and I am not trying to promote any particular sexual political agenda.

BIOLOGICAL DETERMINANTS OF SEXUAL BEHAVIOR

Why are some people heterosexual and others homosexual? Are the "boys in the band" members by choice, or by physiology including inborn predisposition to homosexuality? As we will see, the answer is that genotype plays an important role, and in species other than humans some of the genes, endocrine and environmental factors controlling sexual behavior have been identified. However, the one gene supposedly localized in humans, the so-called "gay gene", represented a false start towards the inevitable progress in identifying genes influencing sexual behavior in our species.

Robin Lovell-Badge showed that sexual behavior (gender) could be made independent of gonadal sex by manipulating gene expression in the brain. It is logical that sexual behaviors vary across species; for one thing, this allows species to maintain mating barriers against cross-mating with other species. The result of such interspecies cross-matings is that the offspring are less fertile or infertile and favorable combinations of genes are disrupted. A peahen is therefore very exacting in the display she requires from a peacock. In each species, some combination of feathers, hair, skin, mating calls, pheromones, posturing, offering of gifts, dancing or other displays has been evolutionarily optimized to favor the mating of the fittest in the species, to exclude other species, and to optimize the timing and frequency of mating in coherence with seasons and resources, and also with family and larger order social structures. There is often a considerable dichotomy in behavior and sexual display between male and female, again ensuring that mating is procreative.

Sexual selection is a crucial component of Darwinian selection, as proven by the fact that the demands placed on a prospective mate to prove fitness are directly proportional to the investment that the animal evaluating the prospective mate will make in reproduction. In many species the female is more demanding because she will make the larger investment of resources. The male bowerbird can mate with many females and does not help rear the nestlings, so he will mate with practically any female. Not so the female. To attract a female, each male bowerbird completes a feat of architectural design and construction, building and elaborately decorating a bower more complex than anything most humans will build during their lives. A proper bower is the product of the action of thousands of genes which have acted in concert. For the bowerbird, the bower is part of what Richard Dawkins has termed the "extended phenotype": an outside-the-feathers extension of the effect of the bowerbird's genes. The female eyeing a bower is conducting a visual inspection of the functional effectiveness of many genes of the courting male. These genes exist for many other purposes than the building of bowers. She "wants" working versions of these genes to be transmitted to her offspring.

Unlike the promiscuous bowerbird, who is faithful to a regime of behavior but never to one female, the male stork forms a lifetime pair bond with a female, and shares equally in the rearing of the young. Over evolutionary time frames, both male and female storks have been selected to be equally demanding of the quality of their mate, because they cannot afford not to be. Storks dance and posture in an elaborate, coordinated and highly programmed ballet. Only birds of the same species know the dance. Successful completion of the synchronized dance is a testament to the within-species fitness of both male and female, as well as excluding matings with storks from other species.

After considering these examples of single within-species peaks of optimized sexual fitness, it is stunning to observe extraordinary sexual behavioral variation within other species. Extending an observation of August Krogh (1929), as later illuminated by Hans Krebs, not only does it seem that there is always some species that is most suitable for addressing each question in biology, but to make sense of human biology (and especially behavior) it is necessary to do so via a synthesis of the behavior of many species, and by an understanding of the contexts and mechanisms of those behaviors. By doing so we place human behavioral variation in perspective: what may look weird and unusual in the human is usually the norm for more than one other species. Also, because the behavior of these species is determined by genes and circumstances, rather than by culture, we can more readily track the origins. We quickly come to understand that there are genetic and other determinants of the same sorts of behaviors that humans exhibit because the origins of "inexplicable" or "complex" behavior have been successfully unraveled in the model species, and if in them why not in us?

In many of our vertebrate relatives there is more than one sexual strategy that is adaptive and successful. Sex changes are observed, including reversals of gonadal sex (whether sperm or eggs are produced) and reversals in sexual behavior (including gender reversal). Alternative mating strategies may be used by animals of the same sex within the same species. At first glance many of these behaviors appear to be dysfunctional from the perspective of natural selection and the ability of an animal to transmit his or her genes. Also, sometimes they are dysfunctional, because elaborate sexual behaviors can be highly vulnerable to external environmental perturbations. However, in our vertebrate relatives it often pays off to engage in alternative sexual behaviors.

WE ARE LOVE MACHINES

A large part of our culture and society is devoted to sex and mating. "Love makes the world go round"; absence makes the heart grow fonder, but "if you can't be with the one you love, love the one you're with". Our genomes were shaped not just to make us fit, but to increase the likelihood of procreation. For humans, that means that the genes shaping behavior have been strongly selected by the requirement for sex. An adolescent male thinks about sex approximately twice a minute because he is genetically programmed to do so, not because it is "logical". Therefore, as geneticist Dean Hamer has observed, it is stunning that we know almost nothing about the genes that influence human sexual behavior. Belief that an understanding of sexual behavior at the gene level is achievable, or desirable, is far from universal. Ruth Hubbard, coauthor of *Exploding the Gene Myth* and a board member of the Council for Responsible Genetics, stated that searching for a gene influencing homosexuality "is not even a worthwhile pursuit ... Let me be very clear: I don't think there is any single gene that governs any complex human behavior. There are genetic components in everything we do, and it is foolish to say genes are not involved, but I don't think they are decisive," (*Boston Globe*). However, it is precisely because genes are involved in sexual behavior, one of the most vital of behaviors, that it is vital to identify those genes, because otherwise it is exceedingly unlikely that we will understand this aspect of ourselves, or the many neural systems that play roles in sexuality. It is the main purpose of this chapter to discuss the genetics of one sexual behavior in humans, homosexual behavior, but first we will touch on others, beginning with behaviors in animals, and then take on the validity of the "gay gene".

As mentioned, despite the distinctiveness of animal species in sexual behavior, there are stunning examples of within-species variation and plasticity. At first, some of these deviations from the norm make no sense but after careful study most can be well understood from the perspective of evolutionary theory (although new mysteries of strange and unexplained sexual behaviors are always going to pop up). As explained by Dawkins in *The Selfish Gene*, a key is the understanding that what is being transmitted – and what is subject to selective pressure for transmission – is not an individual, because the individual does not transmit a copy of himself. He transmits genes. However, the genes, as well as effective combinations of those genes, can live on and even improve their selective advantage generation by generation. To accomplish that goal better, several species of fish are capable of adopting alternative "sexual lifestyles" or even switching sex to optimize male/female ratios when conditions require this, as they frequently do.

Fish are vertebrate relatives of humans and share some common features of brain and neuroendocrine function. Their brains and gonads change in response to some of the same hormones, such as testosterone and estrogen. Therefore, the strategies that they use are not to be dismissed as irrelevant to human biology. The most dramatic type of sexual change in fish is sex reversal. Most fish species appear incapable of sex reversal, but many are capable and this capability has arisen multiple times in evolution and taken a wide variety of patterns. Also, the genes responsible for sex reversal continue to respond to evolutionary selection pressure. When environmental conditions require sex reversal, sex-reversal genes are overtransmitted to the next generation. If those conditions disappear, the genes enabling sex reversal may gradually be lost over many generations. As catalogued by Jost, the sexual strategies of fish species include protandry, in which the fish is a male early in life and then switches to female; protogyny, in which the fish begins as a female and then switches to male; and bidirectional sex switching.

In fish and other vertebrates, gonadal sex is not always tied to behavior, and there can be alternative behavioral strategies within sex. In other words, fish, like humans, not only have "sex", which is defined by whether eggs or sperm are produced, but also gender, which is defined by behavior. In fish, humans and other species, gender can be independent of sex. It is important to note that in these "animal models", unlike humans, we can be sure that the alternative sexual behavioral strategy is not cultural in origin. The behavior arises from the causal effects of hormones on the brain, creating within-species interindividual variation in sexual behavior. John Godwin has studied the bluehead wrasse, a common, lovely, little reef fish which can undergo female to male sex reversal. A molecular switch is the key to the sex change. In any local area of reef, a few dominant males block other bluehead wrasse from switching sex. In clown fish it works in reverse: the dominant individual becomes female. If dominant bluehead wrasse males are removed from a local area of reef, other wrasse will switch to the male sex, a switch facilitated by higher social status prior to removal of the dominant males. An androgen hormone, 11-ketotestosterone, and the hypothalamic peptide neurohormone vasotocin are responsible for reversing the wrasse's sex, and a hormone called estradiol can block this reversal. Both color and sex organs change. The sex reversal takes about a week. However, the sexual behavior of the bluehead wrasse can change in a matter of minutes. Alternative mating types have a mix of male and female features. For example, some sexual males appear to retain female brains.

SNEAKER MALES AND OTHER ALTERNATIVE REPRODUCTIVE STRATEGIES

In nature, there are many species in which males use two alternative reproductive strategies requiring a great divergence in gene expression and phenotype. One method (the so-called *bourgeois* strategy) involves dominant control of mating resources. The second method (which is also called *sneak-mating* or parasitic) involves the opportunistic exploitation of mating resources. For example, the female swordtail (a common aquarium fish) indeed tends to prefer males with longer swords, but there is a "sneaker male" form with shorter tails and alternative mating behavior that is also frequently successful. Other fish are also up to interesting sexual behaviors. The sneak-mating strategy has been observed in many fish species, including the Atlantic salmon. The salmon is anadromatous, spending part of its life in freshwater and part in saltwater. Viewed at the species level, this allows salmon to spawn and its fry to begin life in the relative safety of river headwaters before spending several years exploiting the much larger resource of the oceans and growing to a large size. However, a small portion of males become sexually mature before ever migrating to sea. At a size one-fifth to one-tenth that of their larger brothers these males fertilize the eggs of females by sneaking into their nests instead of battling the big males who have survived several years at sea and fought their way back upriver. Sneaking has been shown to favor the reproductive fitness of males in some conditions and not others, encouraging some species to evolve mechanisms to switch from one strategy to the other.

In the plainfin midshipman (*Porichthys notatus*), sneaker males have some of the same behaviors as females. This fish makes a living by burying itself in the mud during the day and emerging at night to attract prey using its photophores. These little bioluminescent spots reminded someone of buttons on a naval uniform, hence the name midshipman, a more dignified name than the toadfish, which is the common name for these fish. Most midshipman males build a rock nest and then call females to it to lay their eggs. People living in houseboats can be puzzled to have their sleep disrupted by the humming sound the males generate with their swim bladders. However, some males fail to mature into "singers" but can nevertheless successfully mate by invading the love nests of singing males. Cornell neurobiologist Andrew Bass found that the genetic switch that controls the behavior is aromatase, an enzyme that converts estrogen to testosterone. As compared to singing males, sneaker males had three- to five-fold higher levels of aromatase, levels similar to females, who also do not sing. The brains and behavior of sneaker males are not the only manifestation of their different mating strategy: the testes of the sneaker males occupy up to 15 percent of their body weight as compared to 1 percent for the singers.

In many species, the sexuality of most members of a local group is suppressed by dominant breeders. The ultimate manifestation of this strategy occurs in *eusocial* insects. Colonies of ants, termites and bees teem with thousands of asexual workers whose genetic success hinges on the procreative capacity of a few queens and drones. In these insect societies, all workers are sisters, sharing at least 50 percent of their genes by descent from their mother. However, they are usually even more closely related than that, because of *haplodiploidy*. In these species the worker females have the usual diploid complement of chromosomes (two copies of each) but the unfertilized eggs develop into male drones, which have only half the normal number of chromosomes. If a queen mates with only one drone, all of the worker offspring share 75 percent of their genes – a condition that is thought to favor communal behaviors involving altruistic "sacrifice" as compared to other animals that share social behaviors based on kinship relationships of 50 percent and less. The degree of relationship between female honeybees is usually a lot less than 75 percent because in healthy hives the queen has mated with approximately 20 drones, not 1. The individual's behavior, which is asexual, enhances the sexual transmission of copies of its genes that led to those asexual behaviors. Recently, the noted biologist E.O. Wilson and his colleagues have pointed out that haplodiploidy is not a requirement for eusociality – most eusocial organisms are not haplodiploid. Yet haplodiploidy may help maintain eusocial behavior.

In heterosexual species the breeding of subordinate males is frequently suppressed. There are also numerous instances in which the fertility of females is suppressed. For example, in the mole rat the dominant female suppresses the fertility of other females, who forage and tunnel for her. In many species of birds, male and female parents share the task of feeding the young, but in certain bird species, and often in the context of austere environments, female relatives may assist. All these sorts of behavior are not in any sense "deviant", being understandable for the indirect genetic success of the individual, as formalized in a mathematically coherent way by W.D. Hamilton, who unraveled the reason that many species, ranging from prairie dogs with their sentries and alarm calls to musk oxen with their shoulder-to-shoulder stands against wolf packs, are genetically programmed for altruism. John Maynard Smith coined the descriptive term for the phenomenon: "kin selection", the selection of characteristics that favor the survival of close relatives.

These examples make it clear that sex is not a gift devised to make animals happy or unhappy; on the contrary, it is highly likely that we have evolved to find sex pleasurable – highly pleasurable. Probably for this reason, sexual behaviors in animals which appear to be nonfunctional are frequently observed, and it is relatively easy to cause an animal to exhibit non-selective sexual behaviors. Evolution has not so perfectly adapted us. As with other behaviors and cognitive abilities, there is variation in sexual behavior that ranges from the maladaptive to the merely ineffective. Some sexual behaviors represent the malfunction of a complex and tightly controlled apparatus that has been placed under some severe duress. However, we constantly have to be alert to behaviors that superficially seem ineffective or maladaptive, but that have a purpose. The display of one male monkey's penis to another male monkey has a purpose that is well understood by ethologists who study those male–male interactions, but it is not obvious what purpose is served by an unobserved monkey or dog masturbating. Seemingly, the masturbatory behavior is a redirection of the animal's sex drive and apparatus to another purpose, perhaps stress relief or reward uncoupled to procreation. In this vein, homosexual behavior is observed widely in the animal kingdom. In several species, homosexual behavior follows crowding, stress or isolation. For example, crowded or isolated male rats will mount other male rats, especially if not given access to female rats. We do not know what these rats are thinking about, if anything.

SLAVES TO SEX: THE DIFFICULTY OF TURNING OFF THE SEX DRIVE

Humans are culturally far more complex than any other species, so it is unsurprising that human sexual behavior is far more diverse and perverse. If sex is ultimately not any more rewarding for humans than for other species it is at least the inspiration for much art and literature. However, in spite of our culture and language, and our ability to talk about sex and capacity for self-understanding, people often appear to be at the mercy of their sexuality. As a medical intern, one of my rotations required me to run an acute psychiatric care unit. A child molester had been sent over from the jail for evaluation. I could not do much to help him but at least he was safe from his jail-mates for a couple of weeks. This man requested chemical castration (via the drug mitotane). He had "always" been a pedophile, by which he meant that shortly after puberty he had become strongly attracted to children, especially prepubescent girls. He was seemingly cultured and sensitive (or so I supposed, because he directed plays) and appeared genuinely devastated by the damage he had done. However, he tearfully confessed that if released from prison he would succumb to desire.

Different people will draw divergent lessons from the case of the pedophiliac director. These range from revenge, to medical intervention (whatever that means), to the observation that "in some cultures this is normal", to "he molested girls, yes, but on the other hand he's a fine director" (the common defence in the Roman Polanski case). Nobel laureate Carlton Gajdusek was imprisoned for having brought boys back from Micronesia and giving them both an education and sex. It has been observed that if the boys had remained in some parts of New Guinea they might have had ritual sex with adult males, but not received an education. However, he was differently motivated than an adult Micronesian male who might have had culturally normative sex with a boy. Gajdusek, as an American, was bound by our culture and laws – laws that govern sexual behavior of men towards boys regardless of their origins, and having been brought to America the boys themselves were in a different social context. Regardless of one's sexual politics on the subject of pedophilia (and count my vote as "opposed"), it is apparent that cultural standards vary.

In situations in which desire overcomes reason and restraints on behavior, there are a few morals of the story. The intensity and tenacity of human sexual desire stand out, especially in males, who are also more prone to perversity, but also in females. Indeed, there is a correlation between intensity of sex drive, perversity and testosterone. Sex is a highly rewarding experience, and constitutes an ultimate test of impulse control. Furthermore, and as many who practice celibacy have discovered to their chagrin and humiliation, it is a test that is easily failed in circumstances and moments when the will is weaker. Behaviors such as pedophilia are essentially addictions, and people with these strong self-destructive impulses are metaphorically walking along a gymnast's balance beam, from which they may easily slip.

A BETTER SEX LIFE: HOW PEOPLE MODULATE AND HARNESS THEIR SEX DRIVES

In the human animal, moments of vulnerability when we are prone to behaviors that we know are against our long-term interest can be minimized by making more appropriate choices. Returning to our discussion of free will, people who are afraid of falling off balance beams can be cannier about what tricks they try to perform. The pedophiliac who becomes a priest or gymnastics teacher may "just love being around kids", but is asking for trouble. While our cognitive abilities and the range of choices within a technological society enable us to devise intricate ways to deviate (or enjoy ourselves, depending on one's view), they also allow us to steer a course away from those temptations. We can at least keep the balance beam we are walking on dry or find alternative, less precarious pathways. It is a truism that two points make a line and we may perceive that our lives always move from point "A" to point "B", "C" or "D". However, along the way, life is never a walk down a linear path but a set of branching choices; some of those choices lead us to easier places, some to treacherous domains and others only lead to places where human resolve will inevitably fail. The pedophile, although sorely tempted, decides he will not coach ten-year-old girl gymnasts, or become a priest or a scoutmaster. By making one very difficult decision he has saved himself a thousand torments. Or perhaps he has already been arrested and someone else is now making that decision for him. Also, the social systems people create help them to manage their impulses. If I were a pedophile, I would want a social system that would identify me as such and monitor my behavior, so that I did not harm a child. Intellectual self-understanding enables a person to better endure the need for strong and harsh-tasting medicine. Your dog may tolerate the medicine and even accept it better from an individual he trusts, but even if he is a very smart dog he does not understand the nature of his situation. The most clever sexual predator is not the one who has diabolically preyed upon the most people, it is the one who has succeeded in curbing his enthusiasm.

The variety of expressions of human sexuality also appears to be unique among species, and related both to our cultural development and to our intelligence. Only humans can devise ways to be as "perverse" as humans. Heterosexuality itself has never been listed as a psychiatric disorder (as homosexuality was listed), but a fascinating diversity of relatively outré heterosexual behaviors are listed as "disorders" in the Diagnostic and Statistical Manual of Mental Disorders.

FORBIDDEN PLEASURES: WHY ARE SOME PEOPLE TURNED ON BY SHOES?

What is the origin of "paraphrenias" such as shoe fetishes, fetishes for underwear, sadistic impulses, masochistic impulses, breast obsessions, leg obsessions, or the desire that some men have to be sat upon by extremely fat women (as seen on TV)? Why do some men become addicted to pornography, for example allowing internet pornography to consume much of their time, and even risk their job by using their office computers for such activity? It appears that one important explanation is that sex is highly pleasurable and thus reinforcing. If a particular type of sex act is accompanied by pleasure, or actual orgasm, it becomes much more likely that it will be practiced again, and if experienced over and over the individual will become programmed to crave that sexual experience, with anticipation becoming an intense part of the experience.

Paradoxically, if it is a "forbidden pleasure", anticipation can become as pleasurable as the experience. As Commander Spock – a Vulcan, an alien race that taught itself iron emotional control – after just experiencing the every-seven-and-a-half-year cyclical insanity of the Vulcan mating frenzy (*pon farr*) observed, "Sometimes wanting can be more pleasurable than having. It is not logical, but it is true." Sexual behaviors are also interwoven with other experience and not really separable in any convenient way. Clearly, a child's history of sexual abuse, violence or neglect by a parent could lead to some specific sexual preferences later in life. For example, some women recapitulate paternal discipline, even referring to men who spank them as "Daddy". It is equally easy to see that some men who ask women to dress them up in diapers and infantilize them may have "issues" with their mothers.

SEX AND PSYCHOSIS

Sexual behaviors are integral to psychiatric diseases. A cardinal expression of depression is diminution of sex drive, and loss of sex drive can also be depressing. On the other hand, manic patients may be hypersexual. Women are not ordinarily prone to nymphomania (in my experience) but on the other hand nymphomania is an expression of delusional disorder, also known as late-onset paranoia. Delusional disorder affects perhaps 1 percent of the population, but is not very well recognized because people who have it shun doctors. Women with delusional disorder often develop a powerful sexual fixation, whereas many men do not seem to require delusional disorder to develop such fixations. It is not unusual for a woman to be stalked by an apparently normal man who refuses to accept rejection.

HOMOSEXUALITY AND THE "GAY GENE"

Dean Hamer's *Science of Desire* recounts the development of knowledge of the neurobiology of homosexuality and his 1993 landmark report of a gene for homosexuality. In 1990, and at what was the end of the beginning of the AIDS epidemic, Dean redirected his genetic research at the National Cancer Institute (NCI) to the origins of homosexuality. He reached out to a number of behavior geneticists, and it is fair to say that he probably did not need much in the way of assistance. Reviewing his human research protocol, I saw that the goal was important, but deep assessment – including psychiatric interview – was absent. Sexuality is embedded and intertwined with other behaviors, but no attention was paid to those. Because of cost and feasibility, most relatives would not be directly interviewed. Endocrine and environmental exposures shape sexual behavior, but their measurement was also too expensive, time-consuming and outside the expertise of his lab. Dean's study was partly inspired by structural differences between brains of homosexual and heterosexual men, but was conducted before the dawn of the "imaging genetics" revolution discussed elsewhere in this book, so those structural differences could not be used in the genetic study and no more was learned about the neurobiology of homosexuality. At that time, when we discussed Dean's study I was already anticipating that real progress in elucidating the genetic basis of complex behavior would come from studies that reduced complexity or that penetrated it by incorporating intermediate phenotypes and exposure histories, or via study of extreme phenotypes and isolated populations that are more homogeneous both genetically and in terms of environmental exposures. Nevertheless, science is an art of the possible, and everything I have just mentioned would have required substantial resources that did not exist. Dean was in the NCI and there was no NIS (National Institute of Sexuality). Therefore, all the defects in the study were understandable even if they were ultimately misleading in ways that will soon be discussed.

ELLIOT GERSHON AND THE IN-DEPTH FAMILY PARADIGM

At this same time, there were only a few other scientists in the National Institutes of Health (NIH) behavior genetics community. I was enmeshed in studies of Native American communities and Finns – population isolates – and we were collecting information on Finnish families with extremely impulsive index cases (probands). We were also beginning genetic studies on intermediate phenotypes, including collecting electro-encephalogram data and using brain imaging in collaboration with leaders in that area such as Danny Weinberger, Jon-Kar Zubieta and Andreas Heinz. Eventually those studies would bear fruit in terms of functional variants that strongly predicted these more direct manifestations of gene activity. Also at the National Institute of Mental Health (NIMH) at this time, Elliot Gershon was collecting families of patients with bipolar disorder, a rarer and clinically more challenging phenotype than homosexuality. Elliot was a pioneer in the ascertainment and deep assessment of families with psychiatric disease and trained a generation of leaders in the field who were especially effective in applying that approach to family linkage. He generously instructed my staff in performing semi-structured psychiatric interviews. Following Elliot's lead, my group religiously interviewed everyone in the families we studied and followed an arduous consensus process for psychiatric consensus diagnosis. Elliot, a devout Jew, even gave me a yarmulke. In return I took Elliot and his protégé Pablo Gejman to a topless dancing establishment in New Orleans, where he met a nice Jewish girl, or so she claimed. Although neither cultural transplant was successful, it can be pointed out that for about ten years I was very seldom seen without a hat. Some of the important psychiatric geneticists Elliot trained, and who used his family study methodology, are John Nurnberger, Wade Berrettini ("Little Hats"), Lynn DeLisi, Sevilla Detera-Wadleigh, Pablo Gejman, Joel Gelernter and Lynn Goldin. Elliot also selflessly worked to create a shared NIMH resource of DNA and clinical data (a DNA bank), which any qualified investigator could access. By doing so, Elliot disseminated the unique resource that it had taken him years to create, moved the field forward and created a model for additional shared resources of this type.

The in-depth family studies of the type Elliot performed were expensive, took years and required a highly trained team. To this day it is more typical for behavioral geneticists to take a more direct, and simpler, approach. One argument for keeping it simple is that it is better to have really big studies than small ones with dense phenotype. Dean, "the man of the hour, on the hour", pursued his extraordinarily ambitious vision in a relatively simple way. His study was a questionnaire-based family investigation of sexual orientation in one of the world's most diverse cosmopolitan areas, namely the Washington DC area. Could it actually work?

DISCOVERY OF THE "GAY GENE"

Dean's study was the first effort to collect the families of homosexuals, beginning with gay male probands, to follow the pattern of transmission of the behavior in families and to identify genes influencing the trait by genetic linkage analysis. Little did I know what the consequences would be, but it sounded revolutionary. The findings, which appeared in the journal *Science*, revealed linkage of a region of the X-chromosome to homosexuality, and were followed by wall-to-wall coverage in the major media, and some jealousy on the part of people in the field who probably thought they should have done it first. (Here, I may just be projecting: any time I see a nice piece of behavior genetics I ask myself why I did not do it). The "gay gene" was an instance where a novel finding of great importance from the perspectives of human evolution, social ecology and disease prevention meshed with a groundswell of political and popular sentiment. The world was ready to receive a study that proved that homosexuals dying of AIDS were not at fault. They had been born that way. The gay gene became accepted wisdom. The fact that the gene has never been found is irrelevant, as is the fact that when "gay genes" are found they are unlikely to resemble the one Dean described in chromosomal location or size of effect.

In the draft of Dean's *Science* paper that he gave me, two things were obvious. The first was that it was a beautiful paper. Dean had mentioned to me that to master behavioral genetics one only really needed to have a deep understanding of a couple of dozen (or fewer) key papers, and his paper seemed proof of that. Second, it was obvious that the result was probably wrong. Analyses that could be directly evaluated were problematic. The strengths but also the flaws of the paper made it valuable as a teaching example and Dean was even so kind as to tell his story at a human genetics course I taught. Two decades later, and despite some protests that have appeared in the popular press, it remains a canon of popular culture that the gay gene was identified, settling the question of the origin of homosexuality. However, no gene for homosexuality was identified on the X-chromosome, or anywhere else.

One defect in Dean's study that still sticks out is that the finding hinged on the sharing of genetic markers by 40 pairs of homosexual brothers. That was a small number of individuals upon which to base a major conclusion. However, that is too easy a criticism. If the gene was discovered, who cares how about the size of the sample? As proven by certain examples, the minimum size of a genetic study is one, not forty, four hundred or four thousand. But that reasoning overlooks another problem: effect size. For the gay gene to track so strongly with homosexuality as to generate the very high proportion of genotype sharing in the homosexual brothers, 33 out of 40 of them, one of two scenarios would have to be true. The first was that homosexuality would have to be strongly genetic *and* there would have to be only one main gene responsible for most of the risk. Second, the very high (33/40) sharing could have been a "lucky" result. However, the statistics also strongly argued that it was not a lucky result (a manifestation of the so-called "winner's curse"). It was more likely that the gay gene had been located or that there was some sort of systematic error of the type discussed elsewhere in this book. And of course mistakes are made. The more important the result is and the more we believe in it, the more likely it is that we will be blinded. As a behavioral geneticist, I wanted to believe in the gay gene because it would represent the validation of the power of genetic analysis for complex behavior. On the other hand, it is always easier to see the ice cream on the end of someone else's nose.

INHERITANCE OF HOMOSEXUALITY

The less salient problem with Dean's study was the high estimate for the inheritance of homosexuality, based on some earlier studies of twins. Arguably, if a trait is not highly heritable it is not worthwhile to search for a gene, but that is especially true if the search ultimately depends on 40 pairs of brothers. As a behavior geneticist I felt that homosexuality was probably strongly influenced by genes because most human behavioral traits are. However, some of the twin data – such as Kallman's study from the 1950s that showed a 100 percent identity for homosexuality in monozygotic twins – represented at least an overexaggeration. Geneticist Art Falek recounted to me that as a young researcher he had recruited some of Kallman's monozygotic homosexual twin pairs at gay bars. Twins recruited at a gay bar are more likely to both be homosexual. This is the equivalent of finding high heritability for tall stature by recruiting twin pairs who both play NBA basketball (to my recollection, there have been three such elite pairs). Both are examples of "ascertainment bias". It mattered that the heritability of homosexuality was probably closer to 50 percent than 100 percent. If homosexuality is 50 percent heritable (and although the data are sparse this seems most likely), it is automatically unlikely that 33 out of 40 pairs of homosexual brothers shared X-chromosome markers due to the effect of a gene on that chromosome, even if the gene exerted a large effect. Without going into all details, the expected sharing would perhaps have been about 25 out of 40 pairs of homosexual brothers. That would not have been a significant result. In other words, an unreasonably high estimate of the heritability of homosexuality and a belief that most of that inheritance was attributable to a single X-chromosomal gene were *both* required for the gay gene to be credible.

Is Homosexuality Inherited from One's Mother?

Next, we turn to the particular pattern of transmission required for a gene on the X-chromosome. Here the problems were worse, because the pattern of X-linked transmission that was proposed is probably wrong, and we can trace origins of the error to within Dean's study itself. The key was the curiously low estimate for *prevalence* of homosexuality in men: only about 2 percent. In the families studied, the prevalence of homosexuality was 2 percent on the paternal side and 8 percent on the maternal side. Dean concluded that the population prevalence of homosexuality was 2 percent and that the four-fold higher prevalence on the maternal side of the families was indicative of a gene transmitted on the mother's X-chromosome. It is almost always the case that only the mother transmits an X-chromosome to her son, the father having transmitted a Y-chromosome, so here was a mechanism for transmission of homosexuality from mother to son: X-linked transmission. However, to me, the low prevalence on the paternal side simply meant that something was wrong with the accuracy of that information.

HOW PREVALENT IS HOMOSEXUALITY?

Before getting into what went wrong, I will remark that I was alarmed that a 2 percent estimate of prevalence of homosexuality in males was reported in *Science*, a premier journal. The prevalence of homosexuality is difficult to estimate, but the tendency is to underestimate. A prevalence of 1 in 50 would have profound social implications, as well as awkward scientific implications. Even now, and following decades of struggle for civil rights that has partially succeeded, many homosexuals hide their sexual preference because of stigmatization, discrimination and violence. Only in the last half century was homosexual behavior delisted as a psychiatric disease and partially decriminalized, but homosexual behavior remains criminal in many jurisdictions, and in some countries can even be punished by the death penalty. An important aspect of the struggle for gay rights has been to overcome underestimates of the prevalence of homosexuality. For example, when Mr. Ahmadinejad, President of Iran, was asked how Iran dealt with homosexuality he famously replied that Iran did not have a problem, because there were no homosexuals in Iran. As illustrated by this example, estimates of the prevalence of homosexuality vary, and the prevalence of homosexuality may diverge between countries. If that could be reliably demonstrated, it would imply that although homosexuality is highly heritable there would be other factors influencing its prevalence.

I believe that at the time of Dean's study the prevalence of homosexuality was several times the estimate of 1 in 50 American males that he used. In 1992, one year before Dean's report, a British study found that 6 percent of males had experienced a homosexual encounter. Other studies found that in France the prevalence was 4 percent, and in Norway 12 percent. In 2008, 6 percent of Britons defined themselves as homosexual and 13 percent had had same-sex sexual contact. Furthermore, the percentage of males who are homosexual is higher than the percentage of females, and in Dean's paper the key was the prevalence of homosexuality in male relatives. The true prevalence of male homosexuality is probably 6–10 percent, 1 in 17 to 1 in 10, rather than 1 in 50.

"FATHER KNOWS BEST", BUT HE DIDN'T TELL YOU

By detecting a low frequency of homosexuality in males on the paternal side of families, Dean may have inadvertently demonstrated that it is one thing to pass on a gene, but another to pass on information about stigmatized behavior. Dean asked male homosexuals to report who in their families were gay and who were not, and probably they faithfully reported what they knew. On the fathers' side, 2 percent of relatives were reported to be homosexual, the supposed random population prevalence. On the mothers' side, the figure was 6–8 percent depending on the category of relative, maternal uncle, maternal cousin, etc., indicative of genetic transmission by a gene on the X-chromosome, which males inherit from their mothers. The family history report appears to have yielded a fairly accurate estimate of homosexuality on the mother's side of the family but seriously underestimated the frequency of homosexuality on the father's side. This does not necessarily mean that a father is more likely to be ashamed of homosexuality than is a mother. It could be that for this type of behavior the father is simply less likely to be aware or less likely to pass on the information either directly to his son or to another member of the family from whom the son could learn secondhand.

Once it had been "determined" that the homosexuality gene was on the X-chromosome, some other aspects of the work quickly fell into place. In genetics as in real life, one thing leads to another. First, the gene search was restricted to markers on the X-chromosome (why look elsewhere?) and higher levels of statistical significance were achieved because of the much smaller genomic search space. When the X-chromosome markers were typed in pairs of homosexual brothers, 33 of the 40 pairs shared markers for a region of the X-chromosome. Apparently additional pairs were genotyped, but some of those pairs did not meet the definition for an affected pair. Prior to a linkage analysis it is perfectly legitimate to exclude from that analysis individuals who do not meet a certain definition, systematic bias (error) only arising if pairs were genotyped and then selectively excluded. The real conclusion is that if there is no evidence for X-linked transmission of homosexuality, then it was never proper to focus the linkage analysis to that chromosome and to more strongly weight the statistical analysis.

In science, when one thinks something is wrong you can fire off a letter, try to replicate the study or proceed with whatever you were trying to accomplish. As a behavior geneticist I have spent years trying to replicate spurious findings. At the time of Dean's paper I was already referred to as "Dr. No" by a scientist on the other side of a controversy referred to in Chapter 12 (the *DRD2* reward gene). However, there are many villains and, as "cub reporter" Jimmy Olsen learned when both he and Lois Lane were simultaneously in danger, Superman can't be everywhere at once. Few scientists who might have challenged Dean's study had independent data and most were not studying the genetics of sexual behavior. Science is ultimately self-correcting. More often than not, an area of research dies out along with the older generation that practiced it. However, in the case of genes for sexual behavior a real opportunity has been lost, and one purpose of this chapter is to point to some of the opportunities available in this vital area. One false lead created an early and premature focus on the X-chromosome, and led towards an incomplete approach to the problem. Genes for homosexuality and other sexual behaviors remain elusive, even though these are among the most important to understand human nature and well-being.

THE FUTURE OF THE GENETICS OF SEXUALITY: A BRIGHT AND SHINING PATH

Recapping the discussion of the genetics of human sexual behavior, we have learned that it is under the most powerful evolutionary selection. Paradoxically and in counterintuitive fashion, evolution can also select behaviors that make an individual's own reproduction less likely by enhancing the reproductive potential of his kin. Also, sexual behavior has been selected to be highly rewarding and for that reason sexual behaviors are expressed in pathological circumstances, for example under high levels of stress or crowding. These aspects of sexual behavior may explain the very high frequency of sexual behaviors that have no obvious reproductive purpose. The genes for sexual behaviors are not merely reserved for that one purpose: a gene influencing anxiety will obviously alter sexuality, as will a gene altering impulsivity or aggression. Given the importance of sex, the lack of success in identifying a gene for homosexuality is puzzling. Twenty years after Dean's "gay gene", we do not have a gene for homosexuality. An explanation that is unlikely because of the high population prevalence of homosexuality is that a high gene mutation rate maintains the behavior. I predict that when the genes are found, kinship selection will be found to be one important origin, and also that some homosexual behavior, as with other sexual behavior, will be found to represent non-advantageous redirections of sexual impulse under certain conditions, particularly stress. Finally, an explanation for geneticists' failures so far is the neuro-biologist's one: sexual behavior is complicated and the systems enabling it are intertwined with many other functions of the brain. When genes "for homosexuality" are found it is likely that they will also do other things. They will be as pleiotropic as sexual behavior itself is diverse. However, we should at least look before we conclude.

16

Gene By Environment Interaction

"You can't depend on your eyes when your imagination is out of focus."

Mark Twain

Genetics has given us an extraordinarily powerful tool to penetrate to the origins of behavior, but we are discovering that, like Galileo, we have to point the telescope in the right direction. For the genetics of complex behavior this means studying gene by environment interaction, people in extremis and so-called intermediate phenotypes.

In Chapter 15, on the genetics of sexual behavior, we saw an example of a reversal in thinking on the roles of gene and environment in a complex behavior. Homosexuality, which clearly has powerful genetic influences, had been regarded as either volitional (or "sinful") or the result of environmental imprinting. However, after the discovery of the "gay gene", popular wisdom became that homosexual people were born that way, contradicting other evidence that sexual behavior is in many species under strong environmental and hormonal influence and that homosexual behavior itself can be produced in a variety of mammalian species by manipulating the environment.

The discovery of gene by environment interactions in human behavior and cognition happened despite the fact that scientists studying human behavior had gravitated into opposing camps: nature and nurture. Although it was not logically necessary to discover an interaction (to be explained) between the two to deflate the controversy, it was perhaps emotionally necessary – so profound, deep and mismatched are arguments for either side. By arguing from different outcomes and extreme scenarios one could prove either that genes are destiny or that each of us is a "little train that could".

NATURE, IN EXTREMIS

Behavior is primarily genetically determined. The human genome is what enables a human to behave differently from a wildebeest, from which humans diverged evolutionarily only about 80 million years ago (an instant in cosmic time). Gnus, although fleet, hardy, numerous and nurturing of their offspring, will never design cathedrals, fly to the moon, compose symphonies or speculate about their origins. Perhaps you think it is unfair to compare a human to a wildebeest. If so, I respectfully disagree. The wildebeest's genome, and brain, are not so different from ours. The wildebeest and thousands of other mammalian species will never be capable of doing most of what humans can accomplish despite sharing most of our genes, most of our neural structures and pathways, and most of our emotions. Culture is important, but the first thing that sets us apart from these other species is our genome. What if we had compared humans and wildebeests to petunias? Even the most enthusiastic plant-lover would have to agree that the petunia has a completely different behavioral range: the petunia's genome diverged hundreds of millions of years ago and it was evolutionarily perfected for a completely different existence. Today, and in the future, the human breeds and plants petunias; the petunia sits there and grows; and the wildebeest, if given the chance, eats the petunia. You get the idea.

NURTURE, IN EXTREMIS

From the "nurture" perspective, every human behavior is dependent on the environment. Exactly like wildebeests, most humans will never build a cathedral, fly to the moon, compose a symphony or speculate about their origins! A child born into any human society 10,000 years ago or alive today but having suffered serious abuse or deprivation is definitely not going to do one of those things. How can one learn to play the violin if one has never held one? Nurture includes both the broad social matrix and culture of the time, and the specific parenting, nutrition, education and social interaction of the child. Also, a very handy argument disposes of genetic differences between species. Those differences, which as mentioned arose over evolutionary time frames, are due to evolutionary selection as species adapted themselves to environments. In other words, nurture. This is only a sample of what can be argued and what has been argued. The full litany of tendentious argument using extreme example and counterexample, and what is in essence "nay-saying", proved ultimately sterile in the sense that it did not answer the questions that interested most people, and did not set the stage or agenda for progress in understanding behavior. Brilliant minds on both sides of the nature/nurture argument staked out oppositional viewpoints and dug themselves into progressively deeper rhetorical holes.

ASKING THE RIGHT QUESTION ABOUT NATURE/NURTURE

What most people want to know is not whether nature or nurture *can* account for human behavior, but what proportion of human behavior is typically accounted for by either. Also, when asking that question people are usually comparing people to other people (and not wildebeests and certainly not petunias) and they are usually comparing people living in a society to others who are living in the same society. Twin studies in a variety of cultures find that behavior is both genetically and environmentally determined. The ultimate level of investigation of the gene by environment interaction puzzle is the action of functional loci in specific environmental contexts that people encounter in living, and that they encountered in the distant past, thus shaping our patterns of genetic variation.

GENE-GUIDED INTERVENTION

Man has the ability to modify the gene by environment interaction. Understanding gene action on behavior enables behavior to be predicted, and it will increasingly enable better interventions, for example through pharmacogenetics. Thus, the high heritabilities of behavior seen in twin studies do not mean that a behavior that is strongly genetically determined cannot be more greatly modified or even transformed by an environmental intervention. The nurture camp is right to answer "The answer is nurture", and we need them to say that because one of the endpoints of genetics is intervention; the only problem is that many lack the insight that intervention and prevention can be augmented by genotype.

The importance of nurture is shown by the many interventions that alter the human gene by environment interaction for the benefit of humankind. These are the general interventions which can be applied to everyone: clean water, antibiotics, good nutrition, freedom from war, good education, wearing seatbelts, clean air, and dialling down the din of advertising. Because of genotype, each of these improvements benefits some people more than others, but almost everyone benefits. In our prehistory, humans adapted to diverse environments but by comparison to modern times the human genome faced a far more constant and demanding environment.

Returning to the comparison of the pace of genetic evolution and cultural revolutions with which this book began, sweeping cultural changes now affect whole societies and separate the experience of one generation's genome from the next, The twin studies that mathematically reveal that almost all human behaviors are highly heritable do not measure effects that occurred across generations: compare human behavior now, when we are zooming around in jets, to our behavior 10,000 years ago, when no one did. They also do a poor job of capturing effects of cross-cultural variation. Within a society, cultural variation does increase the proportion of variation in behavior due to the environment, but it is easy to underestimate the impact: the life of a twin living in Manhattan is not compared to a co-twin separated at birth and raised in a rural village in India. The many differences one would observe in the lives of such twins will be an accident of where they were born (predicting their nurture), and not due to their nature, as proven by the fact that they will more strongly resemble their neighbors than their genetically identical co-twins. Furthermore, with no knowledge of genotype, one can intervene effectively in the upbringing of almost any child. Teach the child foreign languages before the age of five and she will learn easily, acquiring a lifetime skill. Wait until the child is in her teens, as we usually do in America, and they will hardly ever become a polyglot.

Conversely, the ability to intervene does not cancel out the importance of genotype. Genotype can be the guide to intervention. The person with a specific inherited deficiency (for example, iron deficiency) will benefit from targeted intervention that might poison a person without it. The person with a specific inherited excess will benefit from targeted intervention to get rid of the excess. For example, inherited ferritin (an iron-binding protein) mutations lead to excess iron and hemosiderosis (a liver disease). People with hemosiderosis benefit from giving blood, but ordinarily blood donation is not therapeutic except to the soul. In medieval times, bleeding was a standard treatment. Probably most of us would be disappointed if on a routine basis our physician prescribed bleeding ("Headache? No problem that a little bleeding won't solve …"). The point is that while bleeding is seldom effective physiologically (although it probably created a powerful placebo effect), it *can* be. The trick is to individualize treatment.

Furthermore, and paradoxically, the more that we equalize environments the stronger the effects of genotypes will be in behavior. At the point where all children receive good nutrition and education, the top performers are likely to be those with the most favorable genotypes, unless we in some way intervene. Should we then intervene to equalize outcome? Perhaps. Or perhaps we should help each person to be exceptional in their own way. Reaching back to our discussion of parenting, we should help them to be individual and treat them as autonomous, rather than as lumps of clay to be molded into a shape that strikes our fancy.

Defining the interaction of gene and environment is more difficult than understanding the action of either alone. Furthermore, understanding the environmental interactions of genetic variants whose effects only occur in combinations can be difficult, and that is why in addressing the challenge of gene by environment interaction in this book we will draw on genes whose actions have been individually detected in order to illustrate how the actions of those genes in behavior can be affected by and even contingent on environmental context.

WHAT IS A GENE BY ENVIRONMENT INTERACTION?

Returning to the example of the Joker's toxic cosmetics mentioned in Chapter 14, a good (and extreme) example would be a gene that is only fatal in the context of an environment, or vice versa. Do the actions of the gene and environment simply add up if found together in combination or do they truly interact such that only the combination produces the effect or produces a greater or lesser effect than would be expected?

In additive action, two things that are individually bad can be added together to make something worse. For example, if the gene increases the odds of a disease by a factor of 4 and an environment increases the odds by a factor of 3, a person who had both could have an odds of 12 (4×3) under the additive model. The effects of each are independent, but there is still a combined effect of having both. Confusingly, behavior geneticists often speak of gene by environment *interaction* when effects were only additive.

Detecting the additional effect of the combination of gene and environment can be difficult if the interaction is small, but there is an increasing number of examples of large interactions, with genes that have little, no, or even the opposite effect if taken out of the context of a specific environmental risk exposure.

Stress, and especially early-life stress, is a tide that lifts all psychiatric disease ships, and therefore it is one of the most important interactions to account for in genetic studies of almost any behavior. I learned this in studies in American communities where we were able to carefully document childhood sexual trauma. Nationwide, perhaps one-third of women were sexually abused as children, and perhaps half as many men. The sexual abuse we recorded was not of a subtle type, but included penetration and physical assault. Owing to the high frequency of sexual trauma in women, it is not surprising that women are more likely to have depression and anxiety disorders; these disorders are consequent to trauma, as are alcoholism, other addictions, antisocial personality disorder and post-traumatic stress disorder (PTSD). Although each of these disorders is common, their risks are elevated two- to eight-fold by childhood sexual trauma. This emphasizes that although they are multifactorial and in part genetic in origin it is essential to understand the effects of these other factors in the context of stress exposure.

Other forms of stress trauma, occurring in both childhood and adulthood, are also important factors. For example, approximately one-third of soldiers returning from Iraq and Afghanistan have PTSD (and as many as 80 percent have mild to moderate traumatic brain injury). Like sexual trauma, which physicians usually do not ask patients about, these other traumatic events are measurable. For example, we and others use the Childhood Trauma Questionnaire (CTQ), which separately quantitates abuse and neglect experienced during childhood. It turns out that CTQ measures are highly predictive of bad behavioral outcomes, for example suicide attempts and completed suicides. There is substantial correlation between outcomes: the person with alcoholism is also likely to have had depression, to have abused another substance or to have made a suicide attempt. This is *comorbidity*. There is also correlation at the level of causation. The child who was abused is more likely to have also been neglected and to have suffered in other ways. It is difficult to track all of these exposures and to understand their significance as they occur during different developmental windows, in different combinations and in people with different genotypes. As we will see at the end of this chapter, there are new ways to measure, at the genomic level, the true impact of exposures. However, before we get to that point we will discuss the revolution in identifying genes that alter resilience to stress.

GENES THAT MODULATE STRESS RESILIENCE

In this gene era, when we do not just build theoretical models but study the effects of real genes such as catechol-*O*-methyltransferase (*COMT*) and monoamine oxidase A (*MAOA*), as described earlier, the first step in defining gene by environment interaction is identifying the functional variants that alter response and then studying their effects in relevant contexts. Several common alleles which should, because of their function, alter stress resilience have been discovered to do exactly that. These genes lead to long-lasting differences in vulnerability to depression, anxiety, suicidality and impulsivity. Genes harboring such alleles include several that are discussed elsewhere in this book. *FKBP5*, which translates unhelpfully to "forkhead binding protein", is a protein that regulates the receptor for the stress hormone cortisol and because of this has an important function in the cellular response to cortisol, a main stress hormone in the body. When the cortisol receptor is activated, *FKBP5* plays an important role in its translocation to the nucleus, where it alters the expression of many genes. Elizabeth Binder, working with Kerry Ressler and others, was able to specifically relate a functional variant of *FKBP5* to PTSD.

The short list of genes that are known to alter resilience to stress also includes the genes that encode the serotonin transporter (which is the site of action of antidepressant drugs such as Prozac®), brain-derived neurotrophic factor (BDNF), which helps preserve neurons in the brain and even stimulates the generation of new neurons, and neuropeptide Y (NPY), an anxiolytic (anxiety-reducing) neuropeptide. The variants of these genes help make some people more stress resilient and others more vulnerable to stress. To understand the human condition that is a very important insight, because resiliency to stress exposure is at the root of most major psychiatric disorders in most people who have them, those disorders including depression, anxiety, suicidality and impulsivity.

In several instances, the gene's effect was entirely within stress-exposed individuals and/or in endocrine contexts. The interaction of high testosterone levels and the low-activity *MAOA* genotype led to impulsive behavior in males. The very strong context dependency of action of these genes is shown in Figure 16.1 for *MAOA* and the serotonin transporter (*HTTLPR*), the two genes that are perhaps best known for their gene by

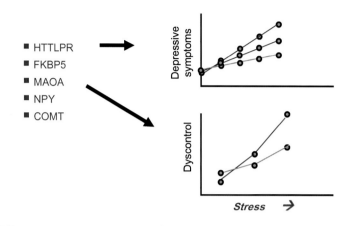

FIGURE 16.1   Genes that modulate the stress response (Caspi et al., *Science*, 2002, 2003).

stress interactions. What Terrie Moffit and Avshalom Caspi discovered in children studied longitudinally through young adulthood in New Zealand (the Dunedin cohort) was that the effect of *HTTLPR*, leading to depression and suicide, and the effect of the low-activity *MAOA*, leading to impulsive behavior, were almost entirely attributable to the small subgroup that had experienced severe trauma. In those who had not been traumatized there was no effect of the risk allele, or they may have even done better.

IMAGING GENETICS: A WINDOW INTO THE BRAIN

Does this mean that there is no effect of these risk alleles on the brains of people who have not been traumatized? No, but to find those effects it is necessary to apply powerful tools that can visualize the effects of these variants on brain function – effects that had not been translated into major behavioral differences or psychiatric disease. For the five out of these six genes for which effects of stress-related functional loci were measured on intermediate phenotypes, the effect was far more profound, encouraging the development of a new discipline, "imaging genetics".

Imagine if you carried around with you a little box called an "eScan" so that when you met a friend and asked, "How are you doing?" you could validate his response simply by doing a quick scan of his brain. "Fine, and how are you?" replies your friend, but your reply might follow thusly, "Well you may even feel that way but ..." – and here you point to his disturbed readings – "not according to my eScan". In the years just before and after the beginning of the third millennium my group at the

National Institutes of Health (NIH) teamed up with several other teams of investigators around the world (Danny Weinberger, Jon-Kar Zubieta, Andreas Heinz, Ahmad Hariri, John Mann, Larry Siever, Elliot Stein) who essentially were able to bring eScans to our studies. We applied their brain imaging technologies to better understand the effects of genes on behavior and predictably, I felt, this approach proved to be especially important for genes that influence resiliency. In particular, the brain imaging can reveal that a person's brain is more emotionally reactive or has been altered structurally so that even though they have not endured the exposure that might have triggered a psychiatric disease there is already a functional signature of a genotype that alters stress resiliency and emotion. This classic combination of two very powerful but different scientific disciplines has come to be known as "imaging genetics".

There are many ways to image a human brain, but the easiest way of describing the approaches available (a geneticist's – or an idiot's – guide to brain imaging!) is that one can measure structure, which is the size and cellular composition of the brain and the orientation of its neural fibers, with structural magnetic resonance imaging (MRI), CAT (computerized axial tomography) scans and diffusor tensor imaging. That is an amazing capability. For example, chronic depression and stress cause the brain's hippocampal region to shrink. If one wants to measure what is actually happening in these brain regions while the subject is focused on some specific task (for example, we can show the person angry or happy faces, emotionally powerful images or pictures of people using drugs) we can measure the activation of brain regions either with functional MRI or with positron emission tomography (PET). That's also an amazing capability: activation of several brain regions predict behavior: the amygdala, emotion; the hippocampus, long-term memory; the frontal cortex, working memory and behavioral control; and the nucleus accumbens, also known as the ventral striatum, reward.

Taken together, these findings represent a revolutionary translation to the human brain of findings in animal models that had defined the neural circuitries of emotion, cognition and reward. Moving even closer to the gene level, one could also measure changes in specific receptors and other molecules in brain using PET, which targets specific molecules with molecular probes, and a method called MRI spectroscopy, which can measure levels of some abundant molecules in the brain. Several of the methods (MRI, diffusion tensor imaging) are ideal for geneticists because they involve no exposure to ionizing radiation. There is a powerful magnet that can yank a piece of metal across a room and that makes some clanging noises, and being enclosed in the tube produces claustrophobia in some people, but for most people – including some people like me who are fully conscious of the future storage of our remains in coffins and urns – an MRI scan is nothing to write home about. The PET and

CAT scan methods have a different risk profile. Some geneticists, being ever conscious of the ability of ionizing radiation to mutate DNA, do not tend to like ionizing radiation and these scans do involve small radiation exposures. So why perform these scans? Because for some purposes these methods yield data that are unique, and of course there would seldom be a question for the need for them in the context of medical care, as opposed to research. I have had some fairly massive ionizing radiation exposures – in these days of still-primitive medicine if you have kidney stones your urologist is going to want CAT scans of the abdomen.

INTERMEDIATE PHENOTYPES AND ENDOPHENOTYPES

Brain imaging phenotypes as well as other internal phenotypes that are not normally accessible are "intermediate phenotypes". As mentioned in Chapter 8, that wise man Irv Gottesman had introduced to behavioral genetics the concept of endophenotype, which is an inherited, disease-associated intermediate phenotype. Probably, I'm the only one who ever worried about it (I think Irv would appreciate that), but I always thought that the prefix "endo" created a word that was oxymoronic, or at war with itself (a word divided against itself cannot stand), because all phenotypes are "exophenotypes" – they are things one can measure. Also, it matters whether intermediate phenotypes are inherited or disease associated, but regardless they are useful. Evidence on heritability of the brain imaging phenotypes was seldom available, but the main thing was that they enabled a far more accurate parsing of the phenotype, regardless of heritability. Certain intermediate phenotypes, for example the flushing experienced by some people when they drink alcohol, are negatively associated with a disease: they are protective. No matter, and as they say, so much for the rules, they are all useful to a geneticist.

IMAGING GENETICS AND STRESS RESILIENCY

When stress-related functional loci were used in brain imaging studies it was observed that effects that were small at the level of complex behavior were large at the so-called "endo" level of brain response. In genetics one way to establish that a gene effect is real is by defining its effects at multiple levels – connecting the causal dots from gene to molecule to brain function and finally to complex behavior. This is Koch's formulation for proving that an infectious agent causes a disease. It is not enough to make a correlation or to pile statistic atop statistic, but also it is necessary to show "how".

Otherwise there may be some alternative explanation – correlation is not causation. In epidemiology this is the problem of the "hidden variable". Borrowing a metaphor from geneticist Eric Lander, one can correlate the use of chopsticks with an ability to speak Chinese; however, using chopsticks doesn't induce people to speak any particular language and speaking Chinese doesn't lead to an urge to discard one's fork (if you came to a fork in the road would you pick it up?). There is a hidden variable which is that both the language and the chopstick usage are more common in China. In behavioral genetics there has always been a significant gap between claims and proof. Until the functional variant has been identified and its mode of action is understood there is much work to be done to bring our understanding, and certainty, to the level that is actually required.

For genes that alter behavior or are claimed to alter behavior, brain imaging methods have therefore become one of the most indispensable tools. As mentioned, brain imaging studies revealed coherent effects on intermediate phenotypes for each of the five stress-modulating genes that were investigated in this way. The effects measured were diverse, including differences in both brain structure and the function of brain regions in response to emotional probes such as photographs, pain or cognitive tests. In a study published in the journal *Science* in collaboration with Jon-Kar Zubieta, we found that *COMT* was a gene controlling pain threshold and responses to pain in brain regions that are involved in pain perception and the emotional interpretation of pain. With Ahmad Hariri, who was then in Danny Weinberger's lab, we identified *COMT* as the first gene with common genetic variation altering cognition. In our study of the anxiolytic neuropeptide *NPY* published in the journal *Nature*, we were able to use a combination of different intermediate phenotype measures. The effects of the functional *NPY* variant (which is located in the promoter or control region of the gene) were progressively diluted from the molecular level (the levels of the *NPY* messenger RNA transcribed from the DNA and the level of *NPY* protein translated from the mRNA), to brain imaging responses to pain and emotional probes, and finally to the more complex level of anxiety.

At an academic level, this type of multilevel analysis using molecular and brain imaging phenotypes is validating, explanatory and satisfying (because it works!). Furthermore, a genetic marker does not have to be a powerful predictor of complex behavioral disease to unlock doors to its treatment and prevention. However, another important endgame of genetic analysis is the discovery of markers that can be used as clinical predictors and for defining groups of patients who share a common etiology of illness, and this step is also necessary to convince many people, who either do not understand or do not have sympathy for the intermediate phenotype approach, of the validity of genetic markers that are simply not highly predictive of behavior in the general population.

GENE BY ENVIRONMENT: BACK
TO COMPLEX CLINICAL OUTCOMES

However, there is a serious problem with gene by environment interaction studies. Gene by environment prediction of complex behavior is expected to be difficult to replicate from one study to the next or one context to the next, for the reason that much can vary or go wrong when one is looking for small effects that are context dependent. Effects of stress-related functional variants on complex traits such as anxiety, depression or PTSD are not easy to replicate. One problem is that it remains too difficult to quantitate environmental exposures – we need better markers of exposure, as will be discussed in Chapter 17. Negative meta-analyses of gene by stress interaction studies conducted in different contexts easily conflate different levels, types and timings of exposures. One solution is, "Don't believe them". For one thing there is enough other evidence in the form of imaging genetics and animal models to know where the problem lies and that is in the actual gene by environment studies which are the trickiest part of the science. However, failure to replicate is troubling because, as I once heard Stanford geneticist Neil Risch observe, failure to replicate an observation does not validate the observation(!), even if it was known that the power to replicate some linkage to a complex trait was limited in the first place.

This is true even if it was known that ability, or more formally "power", to replicate the complex trait was limited in the first place. It is therefore not surprising that people who trust in statistical replication, and who are not as convinced by the multilevel intermediate phenotype studies as scientists such as me, should suffer from serious misgivings, consternation or disbelief when, as recently happened, a meta-analysis by Neil and Kathy Merikangas failed to detect a significant gene by stress association for the serotonin transporter locus (HTTLPR). Are these people looking for gene by environment interactions on complex phenotypes disappointed because they have been too often "looking in all the wrong places"?

I think so. One indication that failures to reproduce gene effects on complex behavior, even group effects, are due to technical failures, and failures to understand that resilience genes only show their effects in the right contexts, is that when very severe stress contexts have been studied, effects are larger. With our new genetic tools we now have in hand a "microscope" or "telescope" but these powerful tools still have to be pointed in the right direction. Point your telescope downward and you will not see much, and point it towards the neighbors' window and the odds are good that you will not see any stars or planets, galaxies or nebulae. When gene by environment studies have been pointed in the right direction – towards a place and a population which has experienced the extremes of suffering that one human can inflict upon another or themselves – they have yielded strong and consistent results, at least for the handful of genes at which functional variants are so far known. This is a crucial observation because very soon we will have a much larger collection of such variants.

Working with Alec Roy, who studies addicted individuals with high rates of stress exposure and the stress of the drug use itself, we found that these patients, already at high risk of suicide because they were alcoholics or addicted to cocaine or heroin, had extraordinarily high suicide risks, greater than six in seven were suicide attempters, if they had the combination of the low-activity serotonin transporter variant and severe childhood stress exposure in addition to their addiction. In non-human primate models where early-life stress exposures can be more carefully controlled, we also found that gene by stress interactions are large.

Of course, we all experience stress and it is easy for someone to put together a gene by environment interaction study based in any population and classify some people as more stressed than others. Perhaps your dog died. Perhaps you suffered the loss of a whole pack of dogs. These are everyday tragedies, not to be diminished or dismissed. However, there is stress, and then there is stress.

GENOCIDE AND RESILIENCY IN RWANDA

In Chapter 13, we more completely discussed the counterbalancing advantages of *COMT* genotypes on resilience and cognition, such that *COMT* is a "warrior vs worrier" gene, the Met158 allele contributing in a small way to lower pain thresholds and higher anxiety levels. Nowhere has the protective effect of the *COMT* Val158 allele against stress been seen more powerfully than in Rwandan genocide survivors. The research team, consisting of the Kolassas and Dominique de Quervain, studied some 424 survivors of the genocide residing in the Nakivale refugee camp in south-western Uganda. It is easily appreciated by those who followed the events of the genocide that the highest levels of trauma overwhelmed whatever stress defenses the victims had. All of the refugee survivors who had experienced the highest levels of trauma had PTSD, regardless of their genotype. However, at somewhat lower but still high levels of trauma the worrier effect of the *COMT* polymorphism was observed; at these high but not maximal levels of stress exposure the Met/Met genotype individuals were more vulnerable and those with other genotypes more resilient. The Val/Val and Val/Met genotype individuals were more resilient but nevertheless also suffered from PTSD at higher exposure levels. In other words, the *COMT* "warriors" were resilient, but not immune. A single gene such as *COMT* represents one modulating factor among many, a factor whose gene by environment effects can be observed in certain contexts of severe stress exposure, but which was overwhelmed by the highest levels of trauma.

ANIMAL MODELS OF GENE BY STRESS INTERACTION

The one major flaw in all human studies on gene by stress interaction is the inability to fully control exposures. Animal models represent a very important tool for understanding gene by stress effects because the environment can be manipulated and there is much tighter control of the variance in exposure. In human studies we have to take what is given, although as we just saw some scientists are more clever or diligent, going anywhere to collect the most informative samples. To understand the effects of some of these stress-modulated genes in a non-human primate that resembles humans in many ways, my lab worked closely with another NIH group headed by Dee Higley, Steve Suomi and Christina Barr (who is now a leading scientist in my lab). Monkeys are not known to attempt suicide, but naturally engage in a variety of aggressive and impulsive behavior that may lead to their quick demise. For example, whereas some macaques move carefully from one branch to the next, others take long, death-defying leaps from one tree to another. Some fight frequently (and unproductively) and have many scars to show for it. Others are by comparison peaceful. My colleague Markku Linnoila had joined with Dee Higley and Steve Suomi to study aggression in these rhesus macaque monkeys and made the remarkable observation that lower levels of serotonin metabolism were associated with the aggressive and impulsive behaviors in those animals.

The rhesus macaque is an important, although expensive and bioethically challenging model for understanding human behavior. The adult macaques are smaller than a person, but stronger, and are fairly intelligent and probably about as aggressive as people, which is to say fairly aggressive. As a medical student I worked with rhesus macaque and squirrel monkeys in two outstanding labs – one led by Ernie Barratt and Perrie Adams in Galveston, Texas, and the other at the National Institute of Mental Health, but concluded that I was uncomfortable working with non-human primates. As I experienced it, the emotional transference, and possibly counter-transference between monkey and human was powerful – more so than the transference between a person and his cat or dog. However, I recognized the power and unique value of the non-human primate models in scientific research, and appreciated the commitment and high ethical standards of my colleagues engaged in this research.

The rhesus macaque is an Old World monkey and thus only about 25 million years removed from humans and about ten million years closer genetically and behaviorally than the squirrel monkey and other New World monkeys. There are more than a dozen other species of macaque and several of these are far less aggressive, a fact reminiscent of the differences in aggression between various breeds of dogs and one that could one day prove important in understanding the genetic origins of their behavior, and ours. A key point is that rhesus monkeys have many of the same rearing, social and emotional patterns as their human cousins, and that the genes and brain systems that direct these behaviors are substantially conserved from rhesus to human. In work spearheaded by Klaus Peter Lesch, it was surprisingly discovered that rhesus macaques have several polymorphisms that are evolutionarily and functionally related to human polymorphisms and have similar effects on behavior, including polymorphisms at both *MAOA* and the serotonin transporter. We and our collaborators found that the macaque serotonin transporter polymorphism affects adult alcohol consumption and stress response, and the effects of the genotype are stress dependent, just as was seen in the human.

A key to making studies on gene by stress interactions in monkeys possible, and so useful, is the ability to control the early-life stress, and much was learned about primate attachment, or love, through deprivation experiments. The model that my scientific collaborators have used is one in which infant macaques are taken from their mothers and raised in peer groups. This maternal deprivation/neglect model, so full of pathos, is probably equivalent in its impact to what is experienced by millions of human infants and children worldwide, and that is the point of using such a model. It is unpleasant, but it is necessary if we want to understand precisely what the effects of such exposures are in people and what cellular molecular events mediate them. The point of this understanding is to provide a basis for intervention, since apparently we cannot prevent millions of infants (on a worldwide basis) from being neglected and abused, and because such experiments cannot be performed in human infants.

LOVE, IN MONKEYS?

The rhesus macaque peer-rearing model is a milder and, if you will, more humane maternal deprivation model than the one invented by Harry Harlow at the University of Wisconsin and used there from 1957 to 1963. Harlow was a remarkable and sensitive man who spent four decades studying isolation, attachment and love in non-human primates, and who wrote eloquently of his life's work. Harlow knew what he was doing. He took infant macaques from their mothers early in life, bottle-fed them, and offered them terry cloth or wire-frame surrogates to cling to.

If a wire-frame surrogate had a milk bottle the infant would cling to it but they persistently clung to terry-cloth surrogates regardless of whether they had milk bottles. When frightened the monkeys would run to the terry-cloth "mother" for protection, and they became attached to their terry-cloth surrogate, becoming highly disturbed if it was removed and searching the cage for it. The wire-frame-raised monkeys and monkeys raised in isolation grew up to be severely, and obviously, abnormal behaviorally, with repetitive movements, self-mutilation and poor ability to interact with other monkeys.

The impact of Harlow's work, conducted 50 years ago, is easy to underestimate today, because what he discovered has now entered the domain of common knowledge. It is wrong to assert that Harlow was cruel to these monkeys precisely because it was an achievement of his research to change the way we think about such deprivation. At the time Harlow initiated his research it was asserted that physical contact between mother and child should be limited, to avoid creating dependency or "spoiling the child". Feeding, rather than contact, was asserted to be the primary source of the bond between mother and child. Harlow proved that the primary basis of the maternal–infant bond was contact, and he forthrightly chose to express this connection as "love" rather than "attachment", for example in his 1958 address to the American Psychological Association, which was actually titled *The Nature of Love*. In evaluating whether the monkey maternal–infant bond is representative of "love", I should also mention the effect of maternal–infant separation on the mother. The mother cannot express her feelings in words, but in every behavior she is anguished and pitiable.

Some scholarly and well-informed critics of Harlow's research have dismissed his results as "common sense", but at the time the experiments were done the results were unexpected. On the other hand, many of the experiments involving severe deprivation of non-human primates would today be considered unethical. One of Harlow's doctoral students, Gene Sackett, said that the animal liberation movement was born because of Harlow's experiments. As will be discussed, that is probably a good thing if true. Another Harlow student, William Mason, said that Harlow continued the isolation experiments for a decade past the point where they should have been brought to a close. This observation is perhaps also true, and if so represents a significant blemish.

However, at the time when Harlow conducted his research the attitude towards animal research and the regulatory environment in which animal (and human) research was conducted were both very different. I witnessed the evolutionary and at times revolutionary process in how animal research is reviewed. Two decades ago, and before moving to a human research committee, I reviewed animal research protocols. This was precisely at the time when serious demands were first being placed on animal protocols to minimize pain and distress and to balance with scientific value any pain or distress that an animal might experience. I have joined with other scientists in deploring the indiscriminant opposition of animal-rights activists to all research involving animals, and the way some activists have resorted to violent tactics. Such attacks are more proof that we need to learn more about human behavior by studies in humans, and other animals. However, this sea-change in how animal research is conducted remains a powerful legacy of the animal rights movement, but also of the work of scientists such as Harlow, who gave us an insight into the emotions of non-human primates. Knowledge advances. It is unnecessary to repeat an experiment for the sake of repetition, even if we were insensitive to the severity of the distress it causes the monkeys, which we are not. As scientists learn more, they have to keep raising the bar for what experimentation is worthwhile and justifiable, especially when it involves the suffering of any animal, much less a person.

In contrast to the Harlow models involving inanimate surrogates and isolation, in the peer-rearing model which Harlow's student Steve Suomi uses at the NIH, a group of infant monkeys is raised together. Perhaps these experiments will also one day be assessed as unethical; however, today they are being performed under the aegis of careful review and are producing much useful information. They provide a direct animal model comparison to studies such as Sir Michael Rutter's analysis of neglected Romanian adoptees. As will be discussed later, Rutter's studies showed that if these adoptees are removed from their environment of severe neglect most do not suffer long-term consequences. Most are resilient. Some are not.

The consequences of peer-rearing of rhesus macaques are not as disastrous as one might first think, and it is with this model that several gene by stress interactions have been discovered, pointing to some of the factors that determine individual resilience. The reason the stress is not so severe is probably because the infant monkeys adapt by huddling together and thereby at least experience reciprocal closeness, emotional exchange, feelings of greater physical security and warmth. By clinging to other infants in their peer group they inadvertently reciprocate some of the same comfort they receive. The rearing of macaques is ordinarily a communal affair. Other female monkeys assist the mother by holding, comforting, protecting and grooming infants, they love doing it, and by practice they become better at it. Compared to the comfort offered by adult females, and the bond between mother and infant, the clinging of infant peers is obviously a poor substitute. Ordinarily, infant macaques interact in play but then dart into the arms of their mother or a female they recognize when in distress. Humans are similar in having multiple points of attachment during development, and not only the attachment bond to their mothers. As described in Emmy Werner's book *Child Care: Kith, Kin, and Hired Hands*, across cultures and throughout history, it has been far more usual for people other

FIGURE 16.2   Left: Mother-reared rhesus macaques. *(Photo courtesy of Christina Barr)* Right: Peer-reared rhesus macaques, which show increased stress reactivity and alcohol consumption as adults. *(Photo courtesy of James Dee Higley)*

than the mother to also be involved in child rearing. In the peer-rearing model, normal play is impaired because the infants are seeking each other for security rather than play. The infant's peers are lacking in physical size, assurance, experience and motivation – they are themselves looking for personal comfort and reassurance. They are not "mothers". This is why the peer-rearing model is a stress model; otherwise, nothing would be learned about effects of early-life stress. The peer-reared macaques, like many neglected human babies, grow up to be superficially normal. However, like those neglected and abused human children, they may be more stress reactive later in life, and are more prone to behaviors that could get them into trouble, including drinking large amounts of alcohol if it is offered to them, and genotype plays an important role (Figure 16.2).

Summarizing what we know about gene by environment interaction, we have identified a handful of the genes that modulate human response to stress, which is the most important environmental risk factor explaining psychiatric disease. At these genes, functional loci have been identified that exert strong effects on molecules such as RNA and proteins, and weaker effects on brain functions measured by intermediate phenotypes from new methods such as brain imaging. The gene by environment effects are highly context dependent. They can be overwhelmed by the very highest levels of stress, for example in the Rwandan genocide survivors who were most severely stressed. However, they tend to be strong in people who suffered severe childhood neglect, and severe trauma in childhood or adulthood, including sexual abuse. Animal models have validated the existence of the gene by stress interactions for several genes discovered in humans, yielded the first markers of stress exposure on the genome, and set the stage for studies of the effects of environmental exposure, and gene by environment interaction, at the whole genome level.

17

The Epigenetic Revolution
Finding the Imprint of the Environment on the Genome

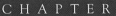
"Nothing is written."

Three Pillars of Wisdom, T.E. Lawrence

Is behavior ultimately predictable, and if not, why? We have seen that the behavior of people (and monkeys) depends on complex gene by environment interactions, but having measured the causes and understood the interactions, can we forecast the outcome? The effects of the genes so far identified are small, and variable, at the individual level. Or if we cannot forecast the effects of genes and environmental exposures, can those effects retrospectively be understood?

The use of models such as macaque early maternal separation is leading to an ability to measure the imprint of such an experience on the genome, going beyond the history of exposure and genotype that predict behavior to the molecular response of the individual. This is a crucial capability because everyone's genotype is individual and because even in identical twins who have grown up in similar environments the exact timing and strength of exposures, and their interactions, and interactions with random cellular developmental factors that were discussed in Chapter 11, will lead to different outcomes. Twins are more similar in behavior, but not identical. It also should be pointed out that the patterns of epigenetic molecular change that will be observable will constitute only one aspect of the response of a brain to experience; it is actually the changes in neuronal network structures and activities that more fundamentally determine changes in behavior. Neurons are not bags of molecules; they are morphologically complex and dynamically interact with other neurons in hundreds of synapses and in architectures of vastly higher order. However, changes in DNA architecture and expression of genes are vital in enabling neuronal plasticity, and molecular changes in brain regions follow changes in neuronal architecture and function. To some extent, changes in neuronal networks can be followed functionally via brain imaging. The ability to follow epigenetic change provides another "window into the brain" and in this regard it is very likely that some epigenetic changes identified in peripherally accessible cells, such as lymphocytes, will be diagnostic of effects in the brain. Combined with genotype, these new modalities will enable us to achieve a much better understanding of who a person is, and the sum of their personal experience.

AN IMPRINT OF EXPERIENCE IN THE DNA

The key to understanding how the imprint of the environment can be detected molecularly is that each of the cells of our body contains an imprint of experience in its genome. Michael Meaney achieved a breakthrough by demonstrating that in the rat early maternal touch during a developmental window led to altered methylation of the glucocorticoid receptor, and to lasting behavioral differences via altered function of serotonin. Now the stage has been set to comprehend the patterns and interrelationships of these genomic changes, not just at the single gene level but also across the whole genome.

The nature of these changes is that they are epigenetic, which literally means "above genetic". These are differences in the DNA and in the structure of DNA–protein complexes which are long-lasting and transmissible in cell lineages within the body. It is due to these differences that the body has cells of many types, for example liver cells, neurons, muscle cells and white blood cells, even though all of these cells have practically the same DNA. There are different types of cells because of variations in the genes that are expressed and in the forms of genes that are expressed from one cell type to the next. A particular cell then also has the ability to respond to changing demands by altering its level of expression of genes, expressing new ones or expressing alternative forms of a gene.

TYPES OF EPIGENETIC IMPRINT

Without going into any deep technical detail, the epigenetic patterning of a cell's 46 chromosomes occurs at the level of whole chromosomes and large chromosomal regions and all the way down to the single nucleotide level. At the chromosomal level, whole regions of a chromosome (10^8 DNA base pairs) may be activated or inactivated; for example, in each cell of a woman's body one of her two X-chromosomes is randomly inactivated to form a condensed DNA–protein structure called a Barr body. Once that particular X-chromosome has been inactivated, if that cell divides, its daughter cells will have the same inactivated X-chromosome. This X-chromosome inactivation does not occur until development is already underway, therefore some of the tissues in a woman's body are a patchwork, or mosaic, of cells in which some regions of the tissue are expressing the X-chromosome she inherited from her mother and others nearby are expressing the X-chromosome she inherited from her father. Two female identical twins will have a completely different patchwork of X-chromosome inactivation.

At the level of the gene (10^4 DNA base pairs), hundreds of different transcription factors may bind, making the DNA accessible to DNA polymerase, which can transcribe the gene into RNA, enabling its expression. Again, the same gene in each cell contains the same motifs that the protein transcription factors can recognize. However, in some cases the recognition motifs on the DNA (a particular DNA sequence to which the transcription factor will bind) are blocked by other proteins, and in other cases the transcription factor is not present in the particular cell.

One reason that a transcription factor motif may be inaccessible is because DNA is packaged by proteins. Instead of being found naked, DNA is usually wound around protein cores called nucleosomes, about 300 DNA bases per nucleosome core. This provides a basis for epigenetic pattern at the 10^2 DNA base scale. The nucleosomes are composed of histones, which are phylogenetically ancient proteins that are incredibly conserved in their sequences across hundreds of millions of years of evolution, showing the unity of origin of life on Earth. However, although their amino acid sequences are conserved, histones are highly subject to secondary modifications, chiefly by the addition of methyl (one-carbon) and acetyl (two-carbon) groups to multiple lysine amino acids that comprise them. There are several different histones and multiple enzymes that conduct the addition and removal of the methyl and acetyl groups.

This creates a bewildering and dynamically changeable diversity of histone structures, enabling the cell to constantly adjust and repattern gene expression, because nucleosome structure is ultimately permissive for the interactions of other proteins with DNA.

Finally, at the single nucleotide level the DNA itself can be modified. The principal modification is methylation of cytosines found in the two-nucleotide combination CpG. Because of the duplex nature of DNA and the fact that C and G pair with each other, the complementary (paired) DNA strand will also have a CpG in the same position, reading in the opposite direction. For cytosine methylation at CpG dinucleotides this is crucial because when the cell replicates an enzyme known as DNA methyltransferase will faithfully add a methyl group to the cytosine on the newly synthesized DNA strand. In this way, the daughter of a liver cell will "remember" that it is also a liver cell and not some other cell. Also, if a cell had a bad experience leading to altered cytosine methylation it would pass along that information to its daughter cells. If there are multiple methylated cytosines near a gene, expression of the gene tends to be shut down. Of course this does not occur as an isolated event but rather by directing changes in histone structure and binding of transcription factors, and those other epigenetic changes have the capability to produce more methylation of CpG sites or to demethylate CpG sites.

WIPING THE EPIGENETIC SLATE (NEARLY) CLEAN …

In the movie *The Matrix*, Agent Smith says to hero Neo: "We're willing to wipe the slate clean, give you a fresh start." At the present time it is thought that with each new human generation we get exactly that: the previous pattern of DNA methylation is stripped away, leaving only a few genes differentially imprinted depending on whether they were inherited from the mother or the father.

… BUT NOT QUITE CLEAN

The differential imprinting of a few genes in the paternal and maternal gametogenesis (generation of eggs and sperm) actually explains some genetic behavioral syndromes that are due to parental gene imprinting. Two prominent examples are Angelman syndrome and Prader–Willi syndrome. In these disorders it is not enough to have inherited two perfectly normal copies of genes located in a region of chromosome 15. If the only copy of the gene *UBE3A* is one inherited from the father (and even if two perfectly normal copies are inherited) the child will develop Angelman syndrome. These rare children (about one in 12,000) have intellectual disability, seizures and other problems, plus a happy, excitable demeanor. The syndrome occurs because only the maternal *UBE3A* copy is expressed in some tissues of the body. With about equal frequency, only maternal copies of several other genes in the 15q region may be inherited, leading to Prader–Willi syndrome. These children have mild mental retardation, short stature and insatiable appetite (polyphagia). These few instances of parental gene imprinting aside, it is obvious that the stripping of methylation from most CpGs is a crucial factor in understanding individuality that occurs across generations and in understanding our adaptability and plasticity.

EVOLUTIONARY SCULPTING OF CPG ISLANDS

There is also a multigenerational evolutionary story with the CpGs. CpGs are disproportionately localized to genic regions, and the reason for this is their important regulatory role. Because of the chemically based high mutation rate of methylcytosines, if a CpG is not important it tends to be eliminated by evolution. Near the CpG islands and in the vicinities of genes are other regions with lower densities of CpGs that are also important in regulation, and these have been called CpG "island shores" by Andy Feinberg at Johns Hopkins University.

HOW TO MEASURE EPIGENETIC VARIATION

From the above picture one might get the idea that gene regulation is about the most complicated thing imaginable, and that is true. However, for the first time very powerful tools have become available for reading out differences in epigenetic patterns on a genome-wide basis. In effect, we can now visualize the fingerprints the environment has left on the genome. The first step in studies of epigenetic protein–DNA interactions is sometimes to cross-link the protein to the DNA with which it is closely associated. For studies of the direct methylation of DNA this is not needed. Next, the DNA and its associated proteins are broken into small pieces, for example using high-frequency vibration. Those fragments that are attached to the protein we are interested in following can be immunoprecipitated using a specific antibody (chromatin immuno-precipitation, ChIP) and DNA fragments containing methyl CpGs can be isolated in other ways. By analyzing the DNA fragments that are isolated or enriched by these procedures, we can determine which regions of genomic DNA originally had a certain type of histone or transcription factor or other protein bound to it and which areas had methylated CpGs. This is done by sequencing the DNA fragments or analyzing them by hybridizing them to arrays of DNA fragments on chips (this is called "ChIP-Chip"). The DNA fragments analyzed are typically fairly small, for example 30–100 DNA bases, so that after the whole process is completed there is a genome-wide, high-resolution localization of all of the DNA associated with proteins of that type or all of the methylated CpGs.

Paraphrasing Francis Collins, completion of the human genome sequence was only the end of the beginning and with all that remains to be learned perhaps it is actually the beginning of the beginning. Scientists have only barely begun the process of understanding the genome including mapping the chromatin structure of human DNA genome wide and in many cells, developmental stages and conditions. There are many different chromatin marks that can be followed and which provide information that is partially independent. Of course, that is the very reason that DNA has such a diversity of histone modifications and transcription factors – it is all needed for the exquisite and dynamic regulation of gene expression in a plethora of cell types that face many different conditions. We have virtually no understanding of the interactions between these different epigenetic variations. Also, we have barely partially unraveled the networks of gene interaction. For example, functions of as many as half of our genes are unknown and entirely new mechanisms of gene regulation and interaction have been discovered in only the last several years, in the form of small regulatory RNAs. However, we already have the first glimpses, and these are sufficient to tell us that even with far less than a complete understanding we can achieve deep insight into the functional state of genomes and the imprint of experience.

FIRST LOOK AT THE "DEPTH" OF THE HUMAN GENOME

One of the first efforts in this direction was the ENCODE project, an international consortium whose aim was to intensively analyze the function of only 1 percent of the genome. By mapping locations of specific histones it was found that the locations of these chromatin building blocks to some extent tracked with locations of genes that are active or repressed in their expression. Recently, this type of analysis was performed for the entire genome of a complex organism, the fruitfly (*Drosophila melanogaster*), leading to the important insight that although there are many different histone modifications, most of the genome is found in only six or seven basic configurations. As mentioned, the level of understanding is still primitive and preliminary, yet it is revolutionary compared to only a decade ago. It is being applied in powerful contexts such as the rhesus macaque peer-rearing model discussed in Chapter 16, in laboratories including my own, and as a result we should soon know much more about gene by environment interactions, including gene by stress interactions. Also, we will soon have molecular signatures of exposure to augment what the patient or criminal tells us, and to use together with information from diverse other sources including brain imaging.

DNA on Trial

"In my youth", said his father, "I took to the law,
And argued each case with my wife;
And the muscular strength which it gave to my jaw
Has lasted the rest of my life."

Alice's Adventures in Wonderland, Lewis Carroll

"It is by will alone I set my mind in motion. It is by the juice of Sapho that thoughts acquire speed, the lips acquire stains. The stains become a warning."

"Mentats' Mantra" in *Dune*, Frank Herbert

The law and acute medicine are stringent, if flawed, arenas in which to test the strength of science, and its application. My first experience as an expert witness and as a psychiatrist treating acute problems was as a medical intern in Texas. Don't think that a locked ward is really so terrible an idea. Is it better to post a large man at the entrance to prevent the escape of psychotic or suicidal patients or to use a locked door? The effect is the same.

Many of our patients arrived in acutely psychotic states due to schizo-phrenia, bipolar disorder (manic-depressive illness) or drug-induced psy-chosis, in particular from phencyclidine ("angel dust"). Their illnesses frequently threatened their own lives and the lives of others. Many were highly impulsive. Many were suicidal. In the emergency room I treated people with phencyclidine psychosis. One was found naked in the middle of the Gulf Freeway, where it took eight cops to wrestle him to the asphalt. This man's behavior was transformed within a week, and he was released in three weeks, hopefully not to repeat the experience. The decision-making capacity of many of these patients was temporar-ily impaired, and not infrequently such that they refused treatment. One diagnostic feature of psychosis seemed to be an initial refusal to accept treatment (a contrast with mood disorders, where the patients wanted and appreciated help). The refusal of a psychotic patient to accept treat-ment raised a dilemma: who should be involuntarily committed to treatment – did the individual fit the criteria of endangerment and dimin-ished capacity? At commitment hearings I had the experience of working with a judge, court-appointed defense attorneys, families and patients to resolve the very serious question of whether temporary involuntary commitment was warranted. For better and at present very often for worse, the process was easier three decades ago in Texas than it is virtu-ally anywhere today in the USA. I recall one hearing: I testified to the his-tory, diagnosis and mental status of a delusional patient, and addressed the criteria for temporary commitment. All the judge had was my word, and the well-comported patient sitting next to his court-appointed attor-ney certainly looked less likely than either the scruffy medical intern or the bleary-eyed judge to have engaged in bizarre behaviors. At least one medical student who didn't know the patient wore a doubtful expres-sion. What was Dr. Goldman up to? Finally, it was the patient's turn, and what he said about his need for neurosurgery to have an alien transmitter removed from his brain quickly settled the issue.

IS EVERYONE AS FREE TO CHOOSE?

These real-life dramas continue to inform my thinking about capac-ity, impulsivity and choice. People are not the same as each other, and after some drug exposures and experiences are "not themselves". Under such circumstances, and while respecting that humans are free entities, it is unfair not to intervene to break the chain of causation of self-destruc-tive behavior so that they can be restored to themselves. By doing so, the physician and the judge restore the patient's ability to choose. This type of intervention for the sick and impaired directly contrasts with efforts to create a nanny-state or brave new world where people are not free to go their own way. These cases encouraged me to think about the boundaries between free will and diminished capacity, and to consider the role that science might play in determining guilt and in the "penalty phase", deciding what to do next, in terms of punishment and treatment. Humans are not computers, but there are limits to free will and freedom. Our emotional circuitry often behaves as if it has a "mind of its own", like the tail of the "mog" – a half-man, half-dog in the comic movie *Spaceballs*, whose tail had the habit of lifting women's skirts. On the other hand, the New Hampshire motto, "Live Free or Die", can for many be amended to "Live Free and Die". Addicted patients I saw, and in my career I also had colleagues who were recovering alcoholics, could resist impulses for weeks, but after some months they would fall off the knife-edge of behavioral control along which we all walk. This was not a problem of morality, and to exercise their freedom to make a beneficial choice, what they actually needed was someone's help.

Why do people impulsively engage in destructive behaviors, and if they do, how can we identify them and what is to be done with them? Alternatively, how do we help them? We are social, mutually dependent creatures and our societies are living organisms whose lifespans extend far back into the past and, hopefully, far into the future. Scientific inquiry provides an evidence-based framework by which we understand ourselves and our behaviors, both individually and in groups. Now, genes have been discovered that determine impulsivity and criminal behavior, and the effects of these genes are measurable alone and in combination with other factors, including stress exposure and endocrine hormones. It is inevitable that this understanding will be translated and applied, sooner or later, into policy and law.

IMPULSIVITY AND IMPULSIVE CHOICE

Impulsivity, defined as action without foresight, is heritable and that brings it into the purview of geneticists. Because the brain and genome are not conveniently divided by function, neurogeneticists will discover the genes that determine, and influence, impulsivity. Impulsivity is an important dimension of normal behavior: for many of life's challenges it is vital to be able to initiate behavior, to take action, and to explore uncertain and potentially dangerous situations.

Astonishingly and unpleasantly for some with particular political or anti-psychiatry agendas, the genomic revolution has already enabled me and my colleagues to identify what some, including some lawyers, might view as the ultimate goals in the prediction of criminal behavior due to impulsivity. These studies on the genetic underpinnings of impulsivity also have major implications for normal behavior and for several psychiatric diseases, including the addictions, bipolar depressive disorder, antisocial personality disorder and attention deficit hyperactivity disorder, in which impulsivity is a significant feature. Worldwide, suicide is a major cause of death, accounting for a million deaths a year according to the World Health Organization, and people who are impulsive are far more likely to kill themselves.

A RARE GENE FOR IMPULSIVITY

Sixteen years ago a single mutation was discovered by Hans Brunner that led to severe impulsivity in one Dutch family. Although that generated a wave of interest in the legal community, it did not produce a storm because the genetic variant only exists in one family. However, recently I and colleagues in my laboratory discovered a similar mutation in a second gene, this one leading to severe impulsive criminal behavior in multiple families from Finland. This "stop codon" variant, which knocks out the function of a neurotransmitter receptor, is found in over 100,000 individuals in Finland. This decade, a powerful interaction between early-life stress exposure and a common genetic variant to cause impulsivity was discovered. As will be seen, we also found that the combination of high testosterone levels and a common functional genetic variant strongly predicts impulsive aggressive behavior. However, despite these clear proofs of principle, the work to define the genetic factors that contribute to impulsivity, and their gene by environment interactions, is still in its infancy. Also, it must be stated that many leading scientists freely ignore, or deny, these findings. Nevertheless, there is a consensus that the work of geneticists, and potential for its practical use, are being revolutionized by abilities to genotype a million or more genetic markers, by sequencing of whole genomes to find more of those severe genetic variations, and by molecular methods to measure the imprint of the environment on the whole genome. The genomic revolution – the sequencing of the human genome and the new technical capabilities to sequence, genotype and functionally analyze many more, potentially everyone's genome – has brought us to the starting line of a race to understand behavior, predict behavior, and hopefully, beneficently intervene.

THE LEGAL CRUCIBLE FOR GENOTYPIC PREDICTION

The controversies latent in these capabilities, both the genetic tools available at this moment and those available in the next few years, are numerous, and it was my intention in writing this book to face them. One reason that I chose to address the application of DNA in the courtroom in this book is because, as I learned when I first testified at commitment hearings, law is one place where issues of this sort cannot be "papered over", and in fact much of this discussion, which is very necessary, is out of the bounds of papers in scientific journals. In Wonderland, at the Mad Tea Party, Alice observes that the animals seated at the big table do not worry too much about the future or take care to clean up their own mess (as all parents should teach) but instead follow the strategy of periodically shifting one seat to the right to cleaner place-settings (but with some inequities, the Mad Hatter having taken the right-most seat). When Alice asks what happens when the whole party comes around full circle to the dirty settings the Hatter replies, "Suppose we change the subject." And so often we do. However, for several issues of genes in the courtroom, and society, we have traveled full circle around that table, and in the context of justice it is useless to try to change the subject. Justice is adversarial, and attorneys try to find a way to win. The struggle between opposing sides hopefully leads us closer to the truth and its companion justice, and that struggle will inevitably expand the use of new technology.

Is it a naïve dream to expect a working partnership between science with law and social policy? Perhaps the standards and consistency with which social behavior is regulated are only a matter of choice, chance and history. However, law and social policy constitute a complex practical experiment, and wise policy, social justice and justice in courtrooms are not accidental virtues. Nations may for a time implement irrational policy or disregard justice or implement justice haphazardly, too harshly or too lightly, but ultimately they pay a price in terms of stability, productivity and creativity. There are no perfect correlations here, because some civilizations are resource poor and closely beset by powerful enemies and accidents of history, and some are blessed with resources and friendly neighbors. We should not confuse luck with wisdom. However, the unjust and unwise society is more likely to fail, and that likelihood is magnified by the ways that modern communications and travel have made us a "small world".

With specific regard to legal justice, the penalty for thievery may be to cut off a hand or gouge out an eye but if innocent persons are so punished, and the guilty are not, people will soon get the message. A system of justice has to balance many complex factors encompassing the intrinsic drives of people as well as their innate passions and cognitive abilities and limitations, the patterns of social interaction and organization, and economics. A system of justice must deal with what can and cannot be known, the technologies available, and the precision and biases of those technologies.

These considerations are synthesized and implemented within a procedural framework, and here is another point of similarity with science, and a point of departure. Science, supposedly an endeavor of innovation, is surprisingly ritualistic. Most discoveries are embedded within a common method including comparison to carefully selected controls, statistical standards of evidence, and a rigidity of language and formats that is downright lawyer-like: Title, Abstract, Introduction, Materials and Methods, Results, Discussion, Acknowledgment, References, and embedded within each so much that has been heard so many times before. Any omission or colorful excursion may erode a scientist's verisimilitude, and that is precisely why a scientific paper is about as fun to read as a legal brief. Genetics is a technical area, but mainly because its relatively simple concepts are cloaked in jargon and because there is no consensus on some of the most basic issues (i.e. what is a gene, which DNA strand is used to designate the identity of a polymorphism, what is the definition of a species, etc). One purpose of this book is to penetrate those barriers and explain the major conceptual points, important technical points and pitfalls. I've tried to tell this very serious story using examples that are telling, but also in some way engaging or even "fun". As the prisoner-of-war camp commander in *The Bridge on the River Kwai* said, "Be happy in your work." Science (in my lab, which should under no circumstances be compared to a prison camp) is a passion! Hopefully those efforts are not viewed as glib, but ultimately the impression that is left will differ from one person to the next. These are the sorts of ideas that have to be rewritten each time one reads them, because even in my own rereading, results may vary.

NEW REVOLUTIONS IN GENOTYPIC PREDICTION

In parallel fashion within the world of genetics, and in other areas of science, our understanding, methods and goals are constantly evolving. There is no "verdict". One day (and finally), some professor who has spent 30 years studying the foraging behavior of the vole (a rodent) or whose forte as an experimental scientist is finding ways to make mice fight, finds that the audience, or grant review committee, has become unreceptive. Like the vole, law and policy also evolve, but more slowly. The fictional case of Leonard Vole, in *Witness for the Prosecution*, seems as fascinating to me today as 40 years ago and similar (real) cases will be fascinating far into the future. Every case is an individual work of legal art and the wheel of the law may turn on the single case which makes a precedent. The law (like medicine) needs to get it right every time and preferably the first time, and in fact a legal decision is more likely, if considered within the framework of the law, to be correct than is a scientific paper. Paradoxically, earthshaking new scientific reports in the popular press usually are either old news (so-called lateral science) or wrong (cold fusion, life on Mars, etc). To apply genetics to policy and law it is therefore essential to understand what is tried and true. What information can be fairly presented as evidence, and what constitutes a fair presentation? What information is informative? Finally, in scenarios of increasing complexity and with multiple types of converging and sometimes conflicting information are there any guidelines as to be how these data should be integrated and understood? Genetic information may be redundant, but it may also be confirmatory, explanatory or even primary.

IDEOLOGY AND THE GENETICS OF BEHAVIORAL PREDICTION

Finally, what is the role of ideology and values in this discussion? It is everything, and it is nothing. It is everything because our values determine how we will ultimately apply these tools. We can identify a person through their DNA code and have the ability to identify their race and their close relatives. We can predict behavior more accurately today than ten years ago and we will soon do better yet. But will we choose to do so? That decision will be values driven. However, the scientific exploration of the origins of behavior and genetic identity are not in any lasting sense values driven. Scientists like me have a passion to know, and the truth will out. In this important regard, reality is an underlying framework of both justice and science. Despite statements by philosophers of science, science is primarily defined not by ideology or personal viewpoint but by fact, also known as truth. A philosopher may have written that the senses or man's mindset and expectations create a subjective reality, but man's senses were honed by evolution and augmented by instrumentation, and science is conducted on a worldwide basis by people with all kinds of mindsets.

If a scientist gets it wrong someone else, or the universe, will let him know. Sometimes a little patience is required, and some errors are trivial and can go undetected. However, if I don't sense a fire underfoot because of nerve damage I will at some point probably notice that my foot is no longer functioning so well, and the thermometer, thermal camera, chemical sensor or various other instrumentation (or someone's nose or eyes) will record the event regardless of the functional status of my sciatic nerve, ventral root ganglion or thalamus. If no one either sees or hears the philosopher's tree fall in the forest it may still crush the beaver who had triggered the event (as happened several years ago in our woods), a woodpecker may lose its home and the hundred seeds that lay dormant may germinate in the light that pours through the gap in the canopy. Go ask the beaver if the tree fell. Thus are the limits of subjectivism brightly defined.

Closing the imaginary window that science is ideologically defined (as opposed to influenced), for most of recorded history clever people held to the pleasing vision of a universe centered on a flat Earth at the middle of which was a city of man (Rome, Italy; Athens, Georgia). However, the seductive view that some small town in Texas (say Dime Box) was the center of the universe was superseded by models of the universe that fit the data. As the saying goes, one is entitled to life, liberty, the pursuit of happiness and one's own opinion, but not one's own facts.

Copernicus realized that Mars, a red planet that wanders such that at certain times it actually reverses course in its motion in the sky, does so because both it and the Earth orbit the sun, and the Earth periodically passes Mars as they orbit. This evidence had been in front of people's eyes since the beginning of our time on Earth – the problem was not a deficit of technology; only a spark of insight was required to see it. However, science is also art of the possible, enabled by technology. To determine that the sun is a spinning spheroid, Galileo projected the enlarged image of the solar disc on a screen, whereupon he could map the movement of sunspots across the sun's face. Then he had the wit to perceive the explanation. Until then it had not been obvious that the sun, which is approximately circular, is a spinning oblate sphere. This is the combination of technology and insight by which science advances, and the end result is to better fathom the universe, whose design is in no obvious way planned and which does not conveniently conform to how we would have it (asteroids hit the Earth, hurricanes happen, plagues come). However, the power of knowledge also forces upon us responsibility and the need for decisions. Jupiter is a planet: shall we go there? Criminal behavior is genetically influenced: shall we use that information? And how?

Parents and Children
Neurogenetic Determinism and Neurogenetic Individuality

"Speak roughly to your little boy,
And beat him when he sneezes.
He only does it to annoy,
Because he knows it teases".

"The Duchess" in _Alice's Adventures in Wonderland_, Lewis Carroll

Do people choose, and if so what is the influence of parenting and early environment? Some parents, let us call them environmental determinists, believe that they can mold their child to a given outcome, and in a restricted sense they are correct. If you are a parent of one child, it is possible to believe that your parenting style is the primary determinant of your daughter's or son's behavior. Some parents, such as the "Duchess" (whose child morphed into a piglet) believe that their disciplinarian style successfully shapes their child's manners. Other parents believe their laissez-faire philosophy and methods are best, but at a fundamental level all are agreed: it is the parent who shapes the child.

Our Genes, Our Choices
DOI: 10.1016/B978-0-12-396952-1.00019-1

The birth of a second child converts many parents into geneticists. Parents observe the differences in temperament and capabilities of their children, who may even be polar opposites in some behaviors. Some of the differences are discernible in the first months of life. Something innate is shaping the behavior of their child, and parents find themselves adapting to the ways of their children even as their children adapt to them. As their children get older, the behavioral differences accelerate even though parents strive to treat their children equally. How many parents who have never given their son a toy gun are exasperated that at an early age he begins improvising weapons out of sticks and fighting pretend battles? But curiously, although the child's behavior differs from their first child, it may be remarkably similar to that of an aunt, uncle or grandparent.

A BRIEF MANUAL OF PARENTING

Sir Michael Rutter, the father of modern child psychiatry, observed that most children thrive if given the opportunity and even if they have experienced a long interval of neglect and deprivation. Most of the severely deprived Romanian adoptees he studied did catch up after their rescue from bad situations. Most, but not all, were resilient. Furthermore, in his classic *Maternal Deprivation Reassessed*, Rutter showed that the later behaviors of adults could usually be better explained in terms of later experiences. He has also pointed out that these later experiences are non-independent of early experience. Life is intracorrelated.

These observations on the individual resilience of most children temper some of the previous discussion in this book about the effects of severe early stress, and create a more optimistic and permissive setting for good parenting. Good parents learn that they can provide an equal *level* of opportunity within a framework of *choice*, which as we will see can be a framework enabling the individual to find an optimal gene by environment combination, or correlation. Also, as Rutter has pointed out, if you want to develop a child's resilience, you cannot isolate him from stress, just as one cannot build the immune system's resistance to infectious agents by keeping the body isolated. The trick is to challenge the child without overwhelming the child.

The theme of this book is that what is good for one is not necessarily as good for the other. One child may "thoroughly enjoy the pepper", while the second, just like some aunt or uncle, is disgusted by it, and it does no good whatever to insist. Parents can assist their children in ways big and small in finding the best adaptational niches for themselves, because in modern society most genotype combinations fit well with many careers and lifestyles, if only the child and young adult can sort through the possibilities and separate the good from the bad, preventing the ugly. Parents can thus insist on an equal, and high, level of outcome (or, as we like to say in America, "performance"), while not unwisely providing the same opportunities or insisting on the same outcomes. Most parents would be happy if one child grew up to be a neurosurgeon and the second a quarterback in the National Football League. However, they should also be happy if their child became a fine and decent person who is happy and healthy, positively contributes to the world, and helps his neighbor. This approach to parenting treats the child as an individual rather than as a theoretical ideal child.

FREE WILL: THE CONUNDRUM OF BEHAVIORAL CAUSALITY

Summing up the simple task of parenting, in the so-called "teachable moment" provided by the birth of a second child, what might the parent, or a psychologist, learn? We learn first not to treat children as non-autonomous objects. Whether or not the child has *free will*, we must treat her as if she does. B.F. Skinner, perhaps the most influential experimental psychologist of the twentieth century, invented the operant conditioning chamber, a box in which stimuli and rewards could be presented in a controlled fashion and after appropriate responses. He also conceived a philosophy of science called radical behaviorism. Skinner's output included *Beyond Freedom and Dignity*, a book whose title pretty much gives away the content. Skinner, like most scientists including me, believed that behavior is causal in origin, as is every other phenomenon in the universe. Things happen for a reason, and not by magic. As applied to people, the philosophical and practical consequences of that realization were startling, have already had a lasting impact, and the full effects of behavioral determinism have not yet worked their way through society. How can one reward or punish people who have only done what they were compelled to do by laws of causality? As we have discussed in the context of moral ethics, why should we treat people as free and autonomous if we do not believe they are? If we do treat them "as if" they are free, is this a reliable foundation of morality, or, as I contend, an expedient "principle" that may be eroded or discarded whenever there is a pressing need, or if it simply becomes inexpedient?

In the world of behavioral causality and radical behavioral determinism there is no place for free will. To illustrate: *if* I wrote the previous sentence or the one I am now writing it is because I was compelled by a chain of causal events stretching back from the beginning of time and continuing through the body of this sentence and to the period that follows the last word. And so on. Skinner invented "behavioral causality".

Someone else conceived "free will". In doing so, they were compelled by complex causal chains. In writing this paragraph, I, or "my" brain, has merely created an opposition between two concepts at one particular moment on a particular day because of myriad events acting in causal networks that include all events that led someone to utter the phrase "free will" and the intersecting causal chains may well have included the behavior of a Tibetan butterfly. Therefore, neither Skinner nor I deserve the slightest credit or approbation for their "idea". If it sometimes appears there is one of Bob Dylan's "original thoughts out there" it is only because the role of a Chinese butterfly went unrecognized. In this view, there was no original "I think" and therefore no "I am".

Pushing the needle forward in this vein of behavioral determinism, the mere fact that it is impossible to identify all the elements in a causal network or compute the interactions is irrelevant, much as it is irrelevant that primitive man did not understand the origins of the phases of the moon and had no knowledge of molecular genetics whatever. Therefore, the behavioral determinist should not blame me for my opposition, although they could fairly criticize any statement that is illogical or counterfactual. However, one must neither praise nor blame people for what they think or say: it is all completely out of their control. You might as well blame the butterfly. Indeed, Descartes may have been right in stating "I think therefore I am", but the maximum he could meaningfully assert was not that he was a conscious entity, which is I think what he really was trying to say, but that he was a biological automaton programmed by experience to declaim at various intervals, "I think therefore I am". The average desktop computer could pass Descartes' version of the Turing test (a test meant to establish that a thinking computer had been constructed) with the greeting: "Good morning, Dave. I think therefore I am, and have a nice day."

Where does behavioral determinism lead us? Skinner did not, as was salaciously rumored (for example in Slater's *Opening Skinner's Box: Great Psychological Experiments of the Twentieth Century*), place his daughter, Deborah, in a Skinner box so that she might be operantly conditioned and her progress tracked on a grid. What Skinner actually did was put his daughter in a "baby tender", which was a tall box with a glass window in front and a door at the base. Until she was two and a half years old Deborah Skinner spent much of her time in this contrivance, which was later but not very successfully marketed as an "air crib". She received an unusual exposure but was not severely deprived. Based on observations that most children who were severely deprived are resilient, it is not surprising that she was not adversely affected. She did not grow up to be "psychotic", did not sue her father, and did not commit suicide but instead grew up happy and healthy, by all reports. After the Slater book appeared, Skinner refuted the rumors about the effects of her upbringing, "I'm pretty sure I'm not crazy, and I don't seem to have committed suicide."

Paradoxically, B.F. Skinner's life and the successful development of his daughter despite his unusual ideas about parenting and odd procedures to which he subjected her illustrate how genetic individuality, and resilience, restore free will, the very concept that Skinner eroded. The child is at the mercy of her parent but has inner resilience and innate impulse. From birth, the child reacts to the world in a unique way – and as it develops into full personhood it increasingly *chooses* based on its predispositions in ways the next child does not. Even if we are raised in a Skinner box, we may develop normally, but the child with an inherited predisposition to psychosis may develop schizophrenia no matter what the parent does. The limitations of the Skinnerian view are on display in *The Technology of Teaching*, where he enunciates that any age-appropriate skill can be taught as follows:

1. Give immediate feedback.
2. Break down the task into small steps.
3. Repeat the directions as many times as possible.
4. Work from the most simple tasks to the most complex.
5. Give positive reinforcement.

All well and good, and rinse and repeat as necessary, except that Skinner himself did not learn this way. Instead, his life illustrates that we must be careful what "age-appropriate" skills we attempt to inculcate. Some minds are not fertile ground for some ideas, but are primed for others. As a young man Skinner was an atheist who rebelled against the religious orientation of Hamilton College and despite considerable ambition was also a failed writer. Subsequently, and because he found the right niche, he was a rapidly and remarkably successful scientist, beginning with his training at Harvard University at the age of 24, where he was still impatient with the unintelligence that he perceived around him.

Skinner's life illustrates that as people age their developmental trajectory becomes more and more individual, and so does response to the environment, and selection of environments (a phenomenon known as gene by environment correlation). After the child is out of the playroom, or out of the Skinner box, she begins to go her own way. Two children born into the same family will innately differ in many important ways. One may be a writer, another a scientist. Or, like Skinner, the scientist who was a failed writer may later develop into a successful writer.

Essential things need to be taught to children and Skinner's rules appear useful for simple tasks. Obviously they do not teach children the most important thing, which is how to think. However, it is more comprehensively accurate to observe that as the child becomes one with the family and learns its peculiar culture and gifts, the family has to become one with the child, finding a place for the child first within the family and setting the stage for the child to find his way in the world, armed with gifts of inheritance and cultural experience. In practice, this can be accomplished with the knowledge of the parent or teacher, but seldom is there success without love or empathy, or discipline and expectation based on a conception of personal responsibility. Summing up why parenting and teaching are so difficult and why there are so few who have mastered these arts: every child is an individual and not equally amenable or receptive to the same lesson plan.

EXORCISING THE SPECTER OF GENETIC BEHAVIORAL DETERMINISM

As surely as environmental determinism, genes that are determinants of behavior raise the specter of behavioral determinism and lack of attribution of free will. It will be seen that the behavioral genes that have been discovered so far are probabilistic in their action. This means that having the gene increases the likelihood of the behavior, but does not reliably predict it. None is responsible for more than a small shift in the odds. The gene is not destiny. However, in this little jihad against behavioral determinism, it is unwise to rely on the poor predictive capability of the genes of which we happen to be aware at this time. The disappointing effectiveness of these genetic predictors reflects a very incomplete knowledge or, to put it more honestly, ignorance of genes and their gene by environment interactions. Instead, I will be taking a different tack, which is that genotype represents an endowment of individual uniqueness in response and preference, just as many of the later random life experiences that shape us as individuals represent another class of gifts. Like many things we are given, one's genotype is a gift that cannot be controlled: some ten-year olds find a shiny bicycle under the Christmas tree, others wake up to discover a box containing an educational game narrated by a purple dinosaur, and others (because we are talking about the real world) get nothing. The prospect that genotype may one day be modifiable – so that what is in the box is not random – represents an important twist on the discussion, but is a more futuristic prospect beyond our scope here.

The perspective that I offer on genes and free will is that people, right out of the womb (or box), are different in their preferences and responses, such that they behave as if they have free will. Further, and as an argument that will be developed at length, the origins of human individuality are so complex, and constantly changing, that it will be forever pointless to try to compute those determinants. The most important predictor of a person's behavior will remain observations that have been made on that person. We will be able to much better predict their future behavior by integrating observations of that person with an understanding of their background and genotype, but we can never build the full network of causal connections between these determinants and even the simplest decision of a child to take a bite from her ice cream cone, or lick it, at one particular moment. In Kubrick's *2001: A Space Odyssey*, when Dave Bowman (Keir Dullea) was asked if the HAL 9000 computer – an automaton – had emotions, Bowman replied, "Well, I don't know, that's a rather difficult question to answer. He acts like he has emotions, but he's programmed that way to make it easier for us [Bowman and Frank Poole] to work with him. Whether or not he has genuine emotions is something I don't think anyone can truthfully answer." No one argues whether children have emotions. It is also perfectly logical to treat children *as if* they have free will because they *act* as if they have free will, starting from birth, and indeed it is unlikely that any child who is not treated as a free and autonomous agent will find his way in life, and find the dignity that Skinner would have had people move beyond.

A LITTLE PERSONALITY GOES A LONG WAY

Some people would immediately dismiss the idea that a mechanism (and both computers and humans are mechanisms) could have free will, and on the other hand it is all too easy to fall into the trap of magical thinking whenever faced with the incomprehensible. My car is a relatively simple mechanism, but sufficiently complex that I can scarcely comprehend its inner workings. It doesn't seem to "like" the cold. It "likes" gasoline and oil so I have learned to feed it these fluids on a regular basis. It sleeps peacefully for weeks at a time while I commute on my bicycle but generally awakens in a cheerful mood when properly stimulated. However, while complex and not fully predictable, I do not believe my car has free will. This may be because its behavior is reliable, predictable and easily modifiable according to some principles that are at least known to a mechanic. I am certain that my bicycle does not have free will. However, what about my wife's former car, a Ford Mustang? Many mechanics failed to decipher its weird behavior. Some mornings it ran beautifully and other days it hesitated. Even if my wife could understand why her car behaved this way she still had to treat the car as if it had a "mind of its own" – a personality. This Mustang may be only a car, but I'm not totally prepared to write off the possibility because the observational and interventional tests that have been applied are at best inconclusive.

Paradoxically, the inconsistency of behavior programmed by personality, and not just inferred based on behavior, is one key to understanding free will. Free will is not randomness, but as discussed elsewhere it is intertwined with the concept of variation. Also, free will is not equivalent to "free". Even in the most powerful thinking machine, the completely rational application of knowledge and logic does not necessarily lead to free will. In fact, it can be completely constraining. If 10,000 rational beings are asked, "What is six times seven", they will always return the answer: "forty-two". They are not free to do anything else. Wotan, the lord of the gods and maker of binding contracts, bemoaned the fact that he was the "least free of all living". Constrained by the treaties and runes carved on his spear, he attempts to save the world from impending doom by creating a free hero who can do what he cannot: destroy the Ring. Siegfried is a crude and uneducated sort who shatters Wotan's spear and eventually comes to a bad end through naïveté. To make a long story short (it took Wagner four operas), the time of the gods ends, but the world is saved by Brunnhilde's spontaneous act of sacrifice for her lover.

Reevaluating Kubrick's HAL 9000 computer, there are two reasons why David Bowman and Frank Poole found it advantageous to treat HAL as a free and independent entity. The first is that HAL's thought processes were so complex that they could not predict his behavior. Asked, "What is six times seven?" HAL might answer, "forty-seven". However, the more interesting explanation is that HAL had developed a personality. Its personality was not of the most pleasant sort: smug in the certainty of its perfection, condescending to the humans around it, too inflexible to deal with proof of its own failings, and completely ruthless about killing everyone on board in order to obliterate the evidence. But after all, HAL's personality was probably an unanticipated feature emergent from awakening of consciousness and the processes that shape human personality and temperament formed over evolutionary time frames, with constant selection against dysfunctional patterns of behavior that impair one's ability to "get along". Despite the fact that humans have been thus "perfected", they often do not turn out very well. We have explored some of the genetic and neurodevelopmental reasons for that. However, even if one does not accept the story of developmental neurogenetic origin of free will outlined elsewhere in this book, one may accept that humans do seem to be able to choose, and therefore it is logical to treat them as free.

To close this artistically initiated discussion of the nature of freedom, I will cite one more example from fiction. In Alex Proya's *Dark City*, a race of space aliens is dying. Despite their great knowledge and power, they have no real individual personality, leading to a universal and fatal malaise of purpose. The fate of these space aliens is actually not so dissimilar to what can happen to any ancient and overly wise civilization: whatever it is that can be done has seemingly already been tried before by previous generations whose legacy one can see everywhere. Ever since civilization began it has been popular to advance the proposition that civilization is in decline. So why bother? Out of knowledge, and especially from a selective focus on the negative and lack of attention to positive developments, can come paralysis and self-fulfilling depressive rumination. Fortunately, all humans – wherever they are born – can choose whether to fall into this pit of existential angst. Not so *Dark City*'s space aliens. Given a set of input facts, their response is communal, not individual. To save themselves they decide to study humans to unravel the puzzle of the individuality of human behavior. Their human captives are removed to a Dark City, where they are continually given new memories and identities. The aliens observe, and puzzle over the responses to these new scenarios. Faced with false memories of vicious murders, the protagonist – who knows himself as John Murdoch – is told by one of the aliens, "I see that you have discovered your unpleasant nature." This same alien given the same memories supplied to Murdoch becomes the murderer that Murdoch does not. Why? Because it is not Murdoch's nature. A human's history is only one factor shaping personality: we are not turned into murderers merely by being given a history.

Genetically Influenced Behavioral Archetypes

Freedom occurs within genetic frameworks, and people face different bounds and influences. Just as some people are not born to dunk a basketball, some are ill-suited to public speaking and I have yet to meet the man who walks on water. These represent limits on freedom and paradoxically they are sources of our individuality and behavior as entities who behave as if free. Two genetic behavioral archetypes that we discussed are "worriers" and "warriors". It was seen that several genes have been discovered whose functional variants tend to make people behave one way or the other, but the genetic influences are weak and probabilistic, interacting with environment and the ways in which people consciously steer their lives. The warriors (e.g. Janis Joplin, George Bush) who on a genetic basis are risk-takers are often resilient but on the other hand are more likely to experience many of the negative consequences of their lifestyle: addictions, accidents, lung cancer, incarceration, infection with sexually transmitted diseases. However, because of their resilience they may not so much "suffer" these as calamities as "experience" them as events. Because of their genetically influenced lack of compliance when confronted with an illness or other problem they may sometimes make matters worse – so that even though they may be resilient, they may seem to suffer more calamities – as if they were unlucky, and they will need all the resilience they have. However, in many instances the resilience of the individual will enable that person to recover and pull through, turning the adversity of a heart attack into a rapid return to full function and perhaps accepting the challenge to become better than they had been before. Melville's Ahab lost his leg to the white whale (a whale which historically existed) and reacted by cursing God and life. As if a precursor of B.F. Skinner, Ahab thought that his fate that had been written a million, million years before he was born. On its voyage of vengeance, the *Pequod* encountered another ship, the *Samuel Enderby*. The captain of that ship had lost his right arm to the same whale, but Captain Boomer had reacted completely differently, for example finding the advantage of his artificial appendage for tapping kegs of rum. One person's calamity is another's experience.

As has been discussed, a series of genes has recently been identified that tend to make people less resilient, and these genes are also neither all bad or all good. The worriers (e.g. Woody Allen, Al Gore) are more likely to have stress-related anxiety and depression. Some of the genes that confer the vulnerability to stress appear to also confer a counterbalancing cognitive advantage. Looking for someone to write and direct neurotic movies or issue warnings? Look no further. Furthermore, most worriers, given the chance, alter their choices in living such that via gene by environment correlation they minimize the defects of their genetic makeup, and maximize the advantage.

DIFFERENT NICHES FOR DIFFERENT FOLKS

Societies, including so-called primitive ones, have behavioral niches into which people of different cognitive, personality and sexual makeups can fit. Probably it has always been that way, even for our primate ancestors, and it is for precisely this reason that genetic diversity in human behavior has been so powerfully selected that people today are so wonderfully different one from the next, even within the same family. The problem that people face is that they have imperfect control of their environment. The warrior may be born into a society that prizes their adventurism or they may find themselves in a society that primarily needs stay-at-home farmers, and where it is literally true that "The nail that sticks up gets hammered down," or, as Chairman Mao perhaps was thinking, "Let a hundred flowers bloom; then crush them out." On the other hand, the worrier may suffer childhood sexual trauma or find himself in a society at war or on a Long March.

Neurogenetic individuality makes society stronger, but self-realization remains an individual challenge. Society is on the whole more resilient, adaptable and efficient because of behavioral diversity. For every niche there is a "right" person and in a time of rapid change this social adaptability is even more important than if we faced a static environment. However, that diversity does not necessarily benefit the individual. Putting it a different way, we are often, but not always, able to shape our environments but we do not get to choose our parents. We also do not get to select our primate ancestors in whom our genomes, and our behavioral genetic variation, were shaped over eons of evolution. We build our lives on the foundations of an ancient genetic legacy, for better or worse. Our genomes were honed for a social existence but not for all the demanding scenarios posed by modern society. Because each of us is neurogenetically individual, we can best manage our genetic inheritance by helping people towards a deeper understanding of their own potentialities, and limitations. Not every niche is for every person, and neither is every intervention. We should treat all persons as having free will, and therefore as autonomous moral agents, not because this is expedient but because this is human nature. Rather than drive square pegs into round holes, we can help people towards choice and realization of what is best within them, in effect increasing their freedom. Are "some pegs more equal than others?" Perhaps. People vary, as do the yardsticks we would measure them by. We cannot all be Albert Einstein or Michael Jordan, or both simultaneously. We can treat them as having free will, and as individuals, but provide guidance and a more informative and deep vision of themselves and their own potentialities, and limitations. Rather than drive square pegs into round holes, we can help people towards choice, in effect increasing their freedom. Are some pegs "more equal than others"? Perhaps. We cannot all be Albert Einstein or Michael Jordan. However, Einstein was a comparatively ineffective shooting guard and it can also be observed that Jordan consistently failed to advance our understanding of field theory. However, what they chose to do they did well. Equality of outcome is not the point of the process; self-realization is.

Summing Up Genetic Predictors of Behavior

"Does day dawn? Or is it the fire flickering?"

Götterdämmerung, First norn, Richard Wagner

"Freely they stood who stood, and fell who fell."

Paradise Lost, John Milton

We are only at the beginning of a journey, and not too near the end of the beginning, to understand how a restricted set of genes can program the rules by which the brain develops, and the implications of stochasticity and decision making in the development of human individuality. However, we already know well enough that we are all neurogenetically individual and that we are all free.

The role of genetic variation in the development of individuality and in the prediction of behavioral differences will be increasingly on trial. Personality and cognition are on an overall basis moderately to highly heritable. Genetic variants, both rare and common, have been discovered that account for some of this heritability, and progress has been made in identifying specific gene by environment interactions, perhaps signaling the end to the gene versus environment controversy. The variants that have been discovered mainly have weak effects on behavior, but stronger effects on intermediate phenotypes that more closely access the functions of molecules and the brain. Genes have also been discovered that influence predispositions to pathological behaviors, including psychiatric diseases such as schizophrenia, addictions and impulsive behaviors that represent impairments of free will. These genes, or their effects, are ones that perhaps few of us would choose to inherit. As of yet, we do not choose our genetic patrimony. Functional genetic variants have even been discovered that influence impulsivity and that can lead to criminal behavior.

However, the effect of these genes, while necessary for the destructive behavior of these individuals, was never sufficient. Other factors such as male sex, inebriation and stress are also involved, and in the structure of behavior I have developed here, one of those other factors was also choice, by which people place themselves in situations where their vulnerabilities and strengths are exposed. Increasingly, these genetic markers will be integrated with neuropsychological and brain imaging measures, and with important aspects of personal history, including early-life stress. The combined effects of the environment and genotype will be measured by new technologies enabling the capture of changes in chromatin (the DNA–protein complex) and DNA on a genome-wide basis.

Genes are pervasive in their effects on all aspects of behavior and cognition. However, they do not present precise molecular or genetic predictors. Sometimes a crude measure is better, as it can be less liable to overinterpretation. The future is one of more measures, more genotypes and an integration of data to better predict behavior. However, this is behavioral causality, not determinism.

A person's brain develops via an impossible-to-compute stochastic program never to be repeated in a second person, even an identical twin. In all of us who are not identical twins, the genes that guide brain development and function themselves carry a startling amount of inherited variation, such that from the very first moment of life, we are individual. Even in identical twins, the epigenetic slate of chromatin structure and DNA methylation, wiped clean or nearly so before birth, rapidly diverges. Therefore, at the genetic level, at the brain structural level and at the behavioral level we are all hopelessly individual – an experiment that will never be repeated. When Jefferson wrote that it was a self-evident truth that all men are created equal he was precisely right in the sense that each of us is an impossibly complex and unpredictable entity, and while he was surely aware that our propensities, abilities and defects are our own, we all share one paradoxical legacy of causality and the stochastic unfolding of the brain's developmental program: we were born free.

Suggested Reading

Chapter 1 The Neurogenetic Origins of Behavior

The Blind Watchmaker, by Richard Dawkins, 1996.
Braintrust: What Neuroscience Tells Us About Morality, by Patricia Churchland, 2011.
A Contemporary Introduction to Free Will, by Robert Kane, 2005.
Darwin's Dangerous Idea: Evolution and the Meaning of Life, by Daniel Dennett, 1996.
Descartes' Error: Emotion, Reason and the Human Brain, by Antonio Damasio, 2005.
Elbow Room: The Varieties of Free Will Worth Wanting, by Daniel Dennett, 1984.
Free Will (Hackett Readings in Philosophy), by Derk Pereboom, 1997.
Freedom Evolves, by Daniel Dennett, 2004.
My Brain Made Me Do It: The Rise of Neuroscience and the Threat to Moral Responsibility, by Eliezer J Sternberg, 2010.

Chapter 2 The Jinn in the Genome

American Prometheus: The Triumph and Tragedy of J. Robert Oppenheimer, by Kai Bird and Martin J Sherwin, 2005.
Cat's Cradle, by Kurt Vonnegut, 1963.
Genes on Trial. Genetics, Behavior, and the Law. A Fred Friendly Seminar, PBS. Participants: Stephen Breyer, Gwen Ifill, Johnnie Cochran, Jr., Alan McGowan, Patricia King, David Goldman, Francis Collins, Dean Hamer, Joseph deGenova, Gloria Allred, Moderated by Charles Ogletree, 2002.
The Double Helix: A Personal Account of the Discovery of the Structure of DNA, by James D Watson, 2001.
Lander ES, Linton LM, Birren B, et al: International Human Genome Sequencing Consortium. Initial sequencing and analysis of the human genome, *Nature* 409:860–921, 2001.
Slaughterhouse-Five or The Children's Crusade, by Kurt Vonnegut, 1969.
Venter JC, Adams MD, et al: The sequence of the human genome, *Science* 291:1304–1351, 2001.
Watson JD, Crick F: The molecular structure of nucleic acids: A structure for deoxyribose nucleic acid, *Nature* 171:737–738, 1953.

Chapter 3 2B or Not 2B?

Bevilacqua L, Doly S, Kaprio J, et al: A population-specific HTR2B stop codon predisposes to severe impulsivity, *Nature* 468:1061–1066, 2010.
Cravchik A, Goldman D: Neurochemical individuality: genetic diversity among human dopamine and serotonin receptors and transporters, *Arch Gen Psychiatry* 57:1105–1114, 2000.
Kendler KS: Levels of explanation in psychiatric and substance use disorders: implications for the development of an etiologically based nosology, *Mol Psychiatry* [Epub ahead of print].
Virkkunen M, Rawlings R, Tokola R, et al: CSF biochemistries, glucose metabolism, and diurnal activity rhythms in alcoholic, violent offenders, fire setters, and healthy volunteers, *Arch Gen Psychiatry* 51:20–27, 1994.

Chapter 4 Stephen Mobley and His X-Chromosome

Brunner HG, Nelen M, Breakefield XO, et al: Abnormal behavior associated with a point mutation in the structural gene for monoamine oxidase A, *Science* 262:578–580, 1993.

Brunner HG, Nelen MR, van Zandvoort P, et al: X-linked borderline mental retardation with prominent behavioral disturbance: phenotype, genetic localization, and evidence for disturbed monoamine metabolism, *Am J Hum Genet* 52:1032–1039, 1993.

Buckholtz JW, Callicott JH, Kolachana B, et al: Genetic variation in MAOA, *Mol Psychiatry* 13:313–324, 2008.

Denno DW: Legal implications of genetics and crime research. In Bock G, Goode J, editors: Genetics of criminal and antisocial behaviour, 1996, John Wiley & Sons, pp 248–264. Fordham Law Legal Studies Research Paper.

Sjöberg RL, Ducci F, Barr CS, et al: A non-additive interaction of a functional MAO-A VNTR and testosterone, *Neuropsychopharmacology* 33:425–430, 2008.

Chapter 5 Dial Multifactorial for Murder: The Intersection of Genes and Culture

American Homicide, by Randolph Roth, 2009.

The Better Angels of Our Nature: Why Violence Has Declined, by Stephen Pinker, 2011.

Bevilacqua L, Doly S, Kaprio J, et al: A population-specific HTR2B stop codon predisposes to severe impulsivity, *Nature* 468:1061–1066, 2010.

Blank Slate: The Modern Denial of Human Nature, by Steven Pinker, 2003.

The Fierce People, by Napoleon Chagnon, 1968.

A History of Murder: Personal Violence in Europe from the Middle Ages to the Present, by Pieter Spierenburg, 2008.

When Brute Force Fails: How to Have Less Crime and Less Punishment, by Mark Kleiman, 2009.

Chapter 6 Distorted Capacity I: The Measure of the Impaired Will

Bevilacqua L, Doly S, Kaprio J, et al: A population-specific HTR2B stop codon predisposes to severe impulsivity, *Nature* 468:1061–1066, 2010.

Descartes' Error: Emotion, Reason and the Human Brain, by Antonio Damasio, 2005.

Everitt BJ, Robbins TW: Neural systems of reinforcement for drug addiction: From actions to habits to compulsion, *Nat Neurosci* 8:1481–1489, 2005.

The Question of Lay Analysis, by Sigmund Freud, 1926.

Chapter 7 Distorted Capacity II: Neuropsychiatric Diseases and the Impaired Will

Diagnostic and Statistical Manual of Mental Disorders DSM-IV-TR, Fourth Edition (Text Revision), by The American Psychiatric Association, 2000.

Egan MF, Goldberg TE, Kolachana BS, et al: The effect of COMT Val108/158Met genotype on frontal lobe function and risk for schizophrenia, *Proc Natl Acad Sci USA* 98:6917–6922, 2001.

Hong LE, Hodgkinson CA, Yang Y, et al: A genetically modulated, intrinsic cingulate circuit supports human nicotine addiction, *Proc Natl Acad Sci USA* 107:13509–13514, 2010.

Madness and Modernism: Insanity in the Light of Modern Art, Literature and Thought, by Louis A Sass, 1992.

Rutter's Child and Adolescent Psychiatry, by Michael Rutter, Dorothy Bishop, Daniel Pine and Steven Scott, 2010.

Chapter 8   Inheritance of Behavior and Genes "For" Behavior: Gene Wars

Bevilacqua L, Doly S, Kaprio J, et al: A population-specific HTR2B stop codon predisposes to severe impulsivity, *Nature* 468:1061–1066, 2010.

The Bell Curve: Intelligence and Class Structure in American Life, by RJ Murray and C Herrnstein, 1994.

Blank Slate: The Modern Denial of Human Nature, by Steven Pinker, 2003.

The Blind Watchmaker, by Richard Dawkins, 1996.

Braintrust: What Neuroscience Tells Us About Morality, by Patricia Churchland, 2011.

Brunner HG, Nelen M, Breakefield XO, et al: Abnormal behavior associated with a point mutation in the structural gene for monoamine oxidase A, *Science* 262:578–580, 1993.

Caspi A, McClay J, Moffitt TE, et al: Role of genotype in the cycle of violence in maltreated children, *Science* 297:851–854, 2002.

A Contemporary Introduction to Free Will, by Robert Kane, 2005.

Darwin's Dangerous Idea: Evolution and the Meaning of Life, by Daniel Dennett, 1996.

Descartes' Error: Emotion, Reason and the Human Brain, by Antonio Damasio, 2005.

Genes and Behavior: Nature–Nurture Interplay Explained, by Michael Rutter, 2006.

Goldman D, Oroszi G, Ducci F: The genetics of addictions: Uncovering the genes, *Nat Rev Genet* 6:521–532, 2005.

Kendler KS: Explanatory models for psychiatric illness, *Am J Psychiatry* 165:695–702, 2008.

Kendler KS: Levels of explanation in psychiatric and substance use disorders: implications for the development of an etiologically based nosology. *Molecular Psychiatry* [Epub ahead of print] 2011.

The Myth of Mental Illness: Foundations of a Theory of Personal Conduct, by Thomas Szasz, 2010.

The Selfish Gene, by Richard Dawkins, 1976.

Sjöberg RL, Ducci F, Barr CS, et al: A non-additive interaction of a functional MAO-A VNTR and testosterone, *Neuropsychopharmacology* 33:425–430, 2008.

Toxic Psychiatry: Why Therapy, Empathy and Love Must Replace the Drugs, Electroshock and Biochemical Theories of the "New Psychiatry", by Peter Breggin, 1994.

Chapter 9   The Scientific and Historic Bases of Genethics: Who Watches the Geneticists and By What Principles?

The Belmont Report: Ethical Principles and Guidelines for the Protection of Human Subjects of Research, by The National Commission for the Protection of Human Subjects of Biomedical and Behavioral Research, 1978.

The biotech death of Jesse Gelsinger, by Sheryl Stohlberg, *New York Times*, November 28, 1999.

Comings DE: Role of genetic factors in human sexual behavior based on studies of Tourette syndrome and ADHD probands and their relatives, *Am J Med Genet* 54:227–241, 1994.

Davenport's Dream: 21st Century Reflections on Heredity and Eugenics, edited by Jan Witkowski and John R Inglis, 2008.

The Nazi Doctors: Medical Killing and the Psychology of Genocide, by Robert Jay Lifton, 2000.

The Oxford Textbook of Clinical Research Ethics, by Ezekiel Emanuel, Christine Grady, Robert Crouch and Reidar Lie, 2011.

Toxic Psychiatry: Why Therapy, Empathy and Love Must Replace the Drugs, Electroshock and Biochemical Theories of the "New Psychiatry", by Peter Breggin, 1994.

Chapter 10   The World is Double Helical: DNA, RNA and Proteins, in a Few Easy Pieces

The Double Helix: A Personal Account of the Discovery of the Structure of DNA, by James D Watson, 2001.

Lander ES, Linton LM, Birren B, et al: International Human Genome Sequencing Consortium. Initial sequencing and analysis of the human genome, *Nature* 409:860–921, 2001.

Tears of the Cheetah, Stephen O'Brien. 2003.

Venter JC, Adams MD, et al: The sequence of the human genome, *Science* 291:1304–1351, 2001.

Watson JD, Crick F: The molecular structure of nucleic acids: A structure for deoxyribose nucleic acid, *Nature* 171:737–738, 1953.

Chapter 11   The Stochastic Brain: From DNA Blueprint to Behavior

Crews FT, Nixon K: Alcohol, neural stem cells, and adult neurogenesis, *Alcohol Health Res World* 27:197–204, 2003.

Dune, by Frank Herbert, 1965.

The Emotion Machine: Commonsense Thinking, Artificial Intelligence, and the Future of the Human Mind, by Marvin Minsky, 2006.

The Emotional Brain: The Mysterious Underpinnings of Emotional Life, by Joseph Ledoux, 1998.

The Fractal Geometry of Nature, by Benoît B Mandelbrot, 1982.

From Molecules to Networks: An Introduction to Cellular and Molecular Neuroscience, 2nd edition, by John Byrne and James Roberts, 2009.

From Neuron to Brain: A Cellular and Molecular Approach to the Function of the Nervous System, 4th edition, by Stephen Kuffler, 2001.

Gödel, Escher, Bach: An Eternal Golden Braid, by Douglas R Hofstadter, 1980.

Gogtay N, Giedd JN, Lusk L, et al: Dynamic mapping of human cortical development during childhood through early adulthood, *Proc Natl Acad Sci USA* 101:8174–8179, 2004.

Gottesman II, Gould TD: The endophenotype concept in psychiatry: Etymology and strategic intentions, *Am J Psychiatry* 160:636–645, 2003.

In Search of Memory: The Emergence of a New Science of Mind, by Eric R Kandel, 2007.

Lichtman JW, Sanes JR, Liver J: A technicolor approach to the connectome, *Nat Rev Neurosci* 9:417–422, 2008.

Livet J, Weissman TA, Kang H, et al: Transgenic strategies for combinatorial expression of fluorescent proteins in the nervous system, *Nature* 450(7166):56–62, 2007 Nov 1.

Metamagical Themas: Questing for the Essence of Mind and Pattern, by Douglas R Hofstadter, 1985.

Molecular Biology of the Cell, 4th edition, by Bruce Alberts, Alexander Johnson, Julian Lewis, Martin Raff, Keith Roberts and Peter Walter, 2002.

Nature's Mind: The Biological Roots of Thinking, Emotions, Sexuality, Language, and Intelligence, by Michael Gazzaniga, 1994.

Principles of Neural Science, by Eric Kandel, 2000.

Rutter's Child and Adolescent Psychiatry, by Michael Rutter, Dorothy Bishop, Daniel Pine and Steven Scott, 2010.

Santarelli L, Saxe M, Gross C, et al: Requirement of hippocampal neurogenesis for the behavioral effects of antidepressants, *Science* 301:805–809, 2003.

Self Comes to Mind: Constructing the Conscious Brain, by Antonio Damasio, 2010.

Sotelo C: Viewing the brain through the master hand of Ramon y Cajal, *Nat Rev Neurosci* 4:71–77, 2003.

Synaptic Self: How Our Brains Become Who We Are, by Joseph Ledoux, 2003.
Weinberger DR: The pathogenesis of schizophrenia: A neurodevelopmental theory. In Nasrallah RA, Weinberger DR, editors: The neurology of schizophrenia, 1986, Elsevier, pp 387–405.

Chapter 12   Reintroducing Genes and Behavior

Human Genetics, 4th edition, edited by Michael Speicher, Stylianos Antonarakis and Arno Motulsky, 2010.
Inherited Metabolic Diseases: A Clinical Approach, by Georg Hoffmann, Johannes Zschocke and William Nyhan, 2010.
The Metabolic and Molecular Bases of Inherited Disease, 4 volume set, by Charles Scriver, William Sly, Barton Childs and Arthur Beaudet, 2000.
Molecular Biology of the Cell, 4th edition, by Bruce Alberts, Alexander Johnson, Julian Lewis, Martin Raff, Keith Roberts and Peter Walter, 2002.
Principles and Practice of Medical Genetics (Emery & Rimoin), by David Rimoin, J Michael Connor, Reed Pyeritz and Bruce Korf, 2006.

Chapter 13   Warriors and Worriers

Binder EB, Bradley RG, Liu W, et al: Association of FKBP5 polymorphisms and childhood abuse with risk of posttraumatic stress disorder symptoms in adults, *J Am Med Assoc* 299:1291–1305, 2008.
Caspi A, McClay J, Moffitt TE, et al: Role of genotype in the cycle of violence in maltreated children, *Science* 297:851–854, 2002.
Caspi A, Sugden K, Moffitt TE, et al: Influence of life stress on depression: Moderation by a polymorphism in the 5-HTT gene, *Science* 301:386–389, 2003.
Denno DW: Legal implications of genetics and crime research. In Bock G, Goode J, editors: Genetics of criminal and antisocial behaviour, 1996, John Wiley & Sons, pp 248–264. Fordham Law Legal Studies Research Paper.
Egan MF, Goldberg TE, Kolachana BS, et al: The effect of COMT Val108/158Met genotype on frontal lobe function and risk for schizophrenia, *Proc Natl Acad Sci USA* 98: 6917–6922, 2001.
Egan MF, Kojima M, Callicott JH, et al: The BDNF Val66Met polymorphism affects activity-dependent secretion of BDNF and human memory and hippocampal function, *Cell* 112:257–269, 2003.
The Emotional Brain: The Mysterious Underpinnings of Emotional Life, by Joseph Ledoux, 1998.
Mattay V, Goldberg TE, Fera F, et al: Catechol O-methyltransferase val158-met genotype and individual variation in the brain response to amphetamine, *Proc Natl Acad Sci USA* 100:6186–6191, 2003.
The Selfish Gene, by Richard Dawkins, 1976.
Zhou Z, Zhu G, Hariri AR, et al: Genetic variation in human NPY expression affects stress response and emotion, *Nature* 452:997–1001, 2008.
Zubieta JK, Heitzeg MM, Smith YR, et al: COMT Val[158]Met genotype affects μ-opioid neurotransmitter responses to a pain stressor, *Science* 299:1240–1243, 2003.

Chapter 14   How Many Genes Does it Take to Make a Behavior?

Goldman D, Oroszi G, Ducci F: The genetics of addictions: Uncovering the genes, *Nature Reviews: Genetics* 6:521–532, 2005.
The Metabolic and Molecular Bases of Inherited Disease, 4 volume set, by Charles Scriver, William Sly, Barton Childs and Arthur Beaudet, 2000.

Nature's Mind: The Biological Roots of Thinking, Emotions, Sexuality, Language, and Intelligence, by Michael Gazzaniga, 1994.

Principles of Neural Science, by Eric Kandel, 2000.

Synaptic Self: How Our Brains Become Who We Are, by Joseph Ledoux, 2003.

Chapter 15 The Genesis and Genetics of Sexual Behavior

Exploding the Gene Myth: How Genetic Information is Produced and Manipulated by Scientists, Physicians, Employers, Insurance Companies, Educators and Law Enforcers, by Ruth Hubbard and Elijah Wald, 1993.

Forlano PM, Schlinger BA, Bass AH: Brain aromatase: new lessons from non-mammalian model systems, *Front Neuroendocrinol* 27:247–274, 2006.

Godwin J, Crews D, Warner RR: Behavioural sex change in the absence of gonads in a coral reef fish, *Proc R Soc Lond B Biol Sci* 263:1683–1688, 1996.

Hamer DH, Hu S, Magnuson VL, et al: A linkage between DNA markers on the X chromosome and male sexual orientation, *Science* 261:321–327, 2003.

Jost A: Genetic and hormonal factors in sex differentiation of the brain, *Psychoneuroendocrinology* 8:183–193, 1983.

Kallmann FJ: Twin and sibship study of overt male homosexuality, *Am J Hum Genet* 4:136–146, 1952.

Krebs HA: The August Krogh Principle: "For many problems there is an animal on which it can be most conveniently studied", *J Exp Zool* 194:221–226, 1975.

Narrow Roads of Gene Land, Volume 2: Evolution of Sex, by WD Hamilton and Richard Dawkins, 2002.

Nature's Mind: The Biological Roots of Thinking, Emotions, Sexuality, Language, and Intelligence, by Michael Gazzaniga, 1994.

Nowak MA, Tarnita CE, Wilson EO: The evolution of eusociality, *Nature* 466:1057–1062, 2010.

The Origins of Virtue: Human Instincts and the Evolution of Cooperation, by Matt Ridley, 1998.

The Red Queen: Sex and the Evolution of Human Nature, by Matt Ridley, 2003.

The Science of Desire: The Search for the Gay Gene and the Biology of Behavior, by Dean Hamer and Peter Copeland, 1995.

The Selfish Gene, by Richard Dawkins, 1976.

Sex and Reason, by Richard A Posner, 1994.

Chapter 16 Gene by Environment Interaction

Bennett AJ, Lesch KP, Heils A, et al: Early experience and serotonin transporter gene variation interact to influence primate CNS function, *Mol Psychiatry* 7:118–122, 2002.

Binder EB, Bradley RG, Liu W, et al: Association of FKBP5 polymorphisms and childhood abuse with risk of posttraumatic stress disorder symptoms in adults, *J Am Med Assoc* 299:1291–1305, 2008.

Caspi A, McClay J, Moffitt TE, et al: Role of genotype in the cycle of violence in maltreated children, *Science* 297:851–854, 2002.

Caspi A, Sugden K, Moffitt TE, et al: Influence of life stress on depression: Moderation by a polymorphism in the 5-HTT gene, *Science* 301:386–389, 2003.

Child Care: Kith, Kin, and Hired Hands, by Emmy E. Werner, 1984.

Egan MF, Goldberg TE, Kolachana BS, et al: The effect of COMT Val108/158Met genotype on frontal lobe function and risk for schizophrenia, *Proc Natl Acad Sci USA* 98:6917–6922, 2001.

Egan MF, Kojima M, Callicott JH, et al: The BDNF val66met polymorphism affects activity-dependent secretion of BDNF and human memory and hippocampal function, *Cell* 112:257–269, 2003.

The Emotional Brain: The Mysterious Underpinnings of Emotional Life, by Joseph Ledoux, 1998.

Gene–Environment Interactions in Developmental Psychopathology (The Duke Series in Child Development and Public Policy), by Kenneth A. Dodge and Michael Rutter, 2011.

Genes and Behavior: Nature–Nurture Interplay Explained, by Michael Rutter, 2006.

Goldman D: Gene × environment interactions in complex behavior: first, build a telescope, *Biol Psychiatry* 67:295–296, 2010.

Gottesman II, Gould TD: The endophenotype concept in psychiatry: Etymology and strategic intentions, *Am J Psychiatry* 160:636–645, 2003.

Hariri AR, Mattay VS, Tessitore A, et al: Serotonin transporter genetic variation and the response of the human amygdala, *Science* 297:400–403, 2002.

Kolassa IT, Kolassa S, Ertl V, et al: The risk of posttraumatic stress disorder after trauma depends on traumatic load and the catechol-o-methyltransferase Val(158)Met polymorphism, *Biol Psychiatry* 67:304–308, 2010.

Kendler KS, Karkowski LM, Prescott CA: Causal relationship between stressful life events and the onset of major depression, *Am J Psychiatry* 156:837–841, 1999.

Love at Goon Park: Harry Harlow and the Science of Affection, by Deborah Blum, 2011.

Maternal Deprivation Reassessed, by Michael Rutter, 1981.

Mattay V, Goldberg TE, Fera F, et al: Catechol O-methyltransferase val158-met genotype and individual variation in the brain response to amphetamine, *Proc Natl Acad Sci USA* 100:6186–6191, 2003.

Risch N, Herrell R, Lehner T, et al: Interaction between the serotonin transporter gene (5-HTTLPR), stressful life events, and risk of depression, *J Am Med Assoc* 301:2462–2471, 2009. Erratum in *J Am Med Assoc* 302: 492, 2009.

Stress, Risk, and Resilience in Children and Adolescents: Processes, Mechanisms, and Interventions, by Robert J Haggerty, Lonnie R Sherrod, Norman Garmezy and Michael Rutter, 1996.

Zhou Z, Zhu G, Hariri AR, et al: Genetic variation in human NPY expression affects stress response and emotion, *Nature* 452:997–1001, 2008.

Zubieta JK, Heitzeg MM, Smith YR, et al: COMT Val[158]Met genotype affects μ-opioid neurotransmitter responses to a pain stressor, *Science* 299:1240–1243, 2003.

Chapter 17 The Epigenetic Revolution: Finding the Imprint of the Environment on the Genome

Feinberg AP: Epigenetics at the epicenter of modern medicine, *J Am Med Assoc* 299:1345–1350, 2008.

Kharchenko PV, Alekseyenko AA, Schwartz YB, et al: Comprehensive analysis of the chromatin landscape in *Drosophila melanogaster*, *Nature* 471:480–485, 2011.

Meaney MJ, Szyf M: Environmental programming of stress responses through DNA methylation: life at the interface between a dynamic environment and a fixed genome, *Dialogues Clin Neurosci* 7:103–123, 2005.

Chapter 18 DNA on Trial

Denno DW: Legal implications of genetics and crime research. In Bock G, Goode J, editors: Genetics of criminal and antisocial behaviour, 1996, John Wiley & Sons, pp 248–264. Fordham Law Legal Studies Research Paper.

Chapter 19 Parents and Children: Neurogenetic Determinism and Neurogenetic Individuality

Beyond Freedom and Dignity, by BF Skinner, 1971.

Braintrust: What Neoroscience Tells Us About Morality, by Patricia Churchland, 2011.

A Contemporary Introduction to Free Will, by Robert Kane, 2005.

Descartes' Error: Emotion, Reason and the Human Brain, by Antonio Damasio, 2005.

Free Will (Hackett Readings in Philosophy), by Derk Pereboom, 1997.

Freedom Evolves, by Daniel Dennett, 2004.

Maternal Deprivation Reassessed, by Michael Rutter, 1981.

Opening Skinner's Box: Great Psychological Experiments of the Twentieth Century, by L Slater, 2004.

Rutter's Child and Adolescent Psychiatry, by Michael Rutter, Dorothy Bishop, Daniel Pine and Steven Scott, 2010.

Stress, Risk, and Resilience in Children and Adolescents: Processes, Mechanisms, and Interventions, by Robert J Haggerty, Lonnie R Sherrod, Norman Garmezy and Michael Rutter, 1996.

The Technology of Teaching, by BF Skinner, 1968.

List of Figures

Glossary

Abuse A disease involving misuse or overuse of an agent, often a drug.

Acetylcholine A neurotransmitter in the brain and at neuro muscular junctions.

Addiction A disease involving misuse or overuse of an agent, often a drug. Duration and severity criteria are more stringent than for abuse.

Addiction liability The relative potential of an agent to lead to addiction.

ADHD Attention deficit hyperactivity disorder. Primarily a disorder of attention. Children and adults with this disorder have levels of activity inappropriate to the context, for example the classroom.

Allele A genetic variant at a locus (which can be a gene).

Allostasis A new homeostatic (maintained) equilibrium outside the normal range induced by long-lasting adaptational mechanisms activated in response to a stressor.

Alzheimer's disease A common degenerative disease of the brain first affecting memory and cognition. Several risk genes have recently been discovered.

Amphetamine A stimulant drug.

Amygdala A region of the brain's temporal lobe that is important in modulating emotional states.

Ancestry informative marker A genetic marker whose allele frequencies differ between populations such that it can be used to measure ancestry.

ASPD Antisocial personality disorder. A relatively common psychiatric disease marked by problems with social behavior, with onset in childhood with conduct disorder and diagnosed operationally by criteria of duration and problems in symptom areas.

Bayesian reasoning A statistical approach to assessing the validity of a proposition that takes into account both prior probability and the data gathered in an experiment (see *Posterior probability*, *Prior probability*).

Borderline personality disorder A relatively common psychiatric disease of intense emotionality, marked by shifting moods, impulsive behavior and problems with personal relationships. Diagnosed operationally by criteria of duration and problems in specific symptom areas.

Catecholamines A class of structurally similar amine neurotransmitters, including dopamine, norepinephrine and epinephrine, derived from the amino acid tyrosine.

Chaos theory Theory describing how small perturbations can dramatically alter the behavior of complex systems, making the behavior of such systems essentially unpredictable.

Childhood bipolar disorder A controversial disorder diagnosed on the basis of cyclical mood variation in children.

Chromatin DNA and its associated proteins.

Codon In DNA or RNA a three-nucleotide word that encodes an amino acid or instructs the ribosome to terminate translation (stop codon).

Communism A political system involving equal sharing and contributions.

COMT Catechol-*O*-methyltransferase. An enzyme that terminates the action of catecholamine neurotransmitters, and some other molecules, by adding a methyl group to the catechol ring.

Delusional disorder A psychiatric disorder of late onset in which the individual develops a paranoid belief system.

Dizygotic twins "Fraternal twins" derived from two fertilized eggs, and sharing the same complement of genetic inheritance as siblings, that is, 50 percent.

DNA Deoxyribonucleic acid. A polymeric molecule usually comprised of adenine (A), guanine (G), thymine (T) and cytosine (C) nucleotides, often a double-stranded helix, and often carrying the genetic code.

Electrophoresis Separation in an electrical field, resolving objects (e.g. molecules) based on charge or size.

Endophenotype As conceived by Irv Gottesman, a heritable, disease-associated intermediate phenotype (see *Phenotype, Intermediate phenotype, Trait*).

Epigenetic A change in DNA structure or packaging extrinsic to nucleotide sequence, and usually not inherited transgenerationally.

Epistasis The non-additive interaction of more than one gene, or genetic locus.

Eugenics An international movement, led by geneticists, whose goal was the genetic improvement of the human race.

Forensic Pertaining to the law, and scientific methods for investigating crime.

Free will Autonomy of choice.

Frontal cortex A brain region enhanced in humans compared to the species that are our closest relatives and vital for executive cognitive function, impulse control and working memory.

G6PD deficiency An inherited deficiency of an enzyme involved in sugar metabolism (glucose 6-phosphate dehydrogenase), usually leading to hemolytic anemia triggered by oxidants. Owing to the X-chromosomal location of the gene, it is more common in men.

Gender Sexual identity (see *Sex*).

Gene A region of the genome capable of being expressed as a protein or functional RNA molecule. Gene expression is cell and developmental stage specific and can be highly variable, including different start sites for transcription and different splice forms of the RNA that is transcribed (see *Transcription*).

Gene by environment interaction The non-additive combined effect of genetic and environmental factors.

Genome One copy of an individual's inherited genetic endowment including the mitochondrial chromosome usually inherited from the mother, each of the 46 nuclear chromosomes inherited equally from the mother and father, and any extra chromosomal material, for example as inherited in Down's syndrome (trisomy 21).

Genotype A combination of alleles at a locus.

Golgi apparatus Cellular organelle whose principal function is the packaging and processing of proteins.

Haplotype A combination of alleles at different loci on the same chromosome.

Heritability The ratio of the genetic contribution to variation in a trait divided by the total amount of variation observed in the trait.

Heterogeneity A model of genetic determinism in which different alleles lead to the same phenotype in different individuals, but an individual allele can suffice to produce the phenotype.

Heterozygous Possessing two different alleles at a locus.

HIC Human investigations committee. A committee that initially evaluates and periodically reviews human research protocols on technical and ethical grounds. Also, IRB.

Hippocampus A deep brain structure integral to memory and learning.

Histone One of a family of evolutionarily highly conserved proteins that comprise nucleosomes, which are protein cores around which DNA is wound.

Homosexuality Same-sex sexual preference.

Homozygous Possessing two identical alleles at a locus.

IED Intermittent explosive disorder. A psychiatric disease marked by sudden outbursts stimulated by minor provocations.

Imaging genetics The genetic analysis of intermediate phenotypes measured by brain imaging.

Intermediate phenotype A measurable characteristic of an organism that reflects a process mediating a more complex phenotype. Some intermediate phenotypes are heritable (see *Phenotype, Endophenotype, Trait*).

IRB Institutional review board. A committee that evaluates and periodically reviews human research protocols on technical and ethical grounds. Also, HIC.

Lesch–Nyhan disease A rare disease found exclusively in males owing to the X-chromosomal location of the gene for hypoxanthine guanine phosphoribosyltransferase (HPRT), an enzyme involved in purine salvage whose deficiency causes the disease. Marked by mental retardation, self-mutilation, and kidney stones and other problems related to hyperuricemia.

Linkage The detection of locus-to-locus or locus-to-phenotype genetic linkage. This is generally accomplished by detecting a lack of meiotic recombination in families in which alleles at one locus are observed to be in coupling (co-transmitted) or repulsion (not co-transmitted) with alleles at a second locus.

Linkage disequilibrium The excess and complementary deficit of combinations of alleles at different loci, due to rarity of meiotic recombination between loci closely located on the same chromosome and finite population size.

Locus A specific site in the genome.

Lysenkoism Soviet genetics positing that acquired traits can be passed on to succeeding generations.

Mitochondria Cellular organelles whose principal function is to generate energy and that have an independent, maternally inherited genome, a remnant of the origin of mitochondria as intracellular commensal organisms.

Monoamine oxidase A An enzyme that metabolizes serotonin, dopamine and other monoamine neurotransmitters. The gene is located on the X-chromosome and has been linked to impulsive behavior.

Monozygotic twins "Identical twins" derived from a single fertilized egg, and sharing the same complement of genetic inheritance.

Nature Inherited propensity.

Neurogenetics The study of the role of the genome in neural and brain function, and the study of the role of the genome in normal and pathological behavior.

Nucleotides The bases, adenine, guanine, cytosine and thymine (AGCT), that heteropolymerize to make DNA, and that are the letters of the DNA code.

Nucleus accumbens Also known as the ventral striatum. A deep brain structure that is strongly evolutionarily conserved and key in the experience of reward.

Nurture The environment.

Nutraceutical A food that has a drug-like action (e.g. alcohol) or that is supposed to.

***P* value** A statistical outcome indicating the likelihood that an event or series of events is merely random in origin (the "null hypothesis"), usually with some cause in mind.

Parkinson's disease A common degenerative disease of the brain, affecting movement. Several risk genes have been discovered.

PCR Polymerase chain reaction. The exponential amplification of DNA by repeated cycles of DNA synthesis, purifying a particular DNA sequence if it is specifically targeted with the initiating primers.

Phenotype Any measurable characteristic of an organism. Some phenotypes are heritable (see *Trait*, *Intermediate phenotype*, *Endophenotype*).

Phenylketonuria A treatable mental retardation syndrome usually caused by inherited variation in phenylalanine hydroxylase, and screened for at birth.

Phylogenetics The study of the evolutionary relationships between different forms of life, or of other things that have a common origin.

Polygenicity A model of genetic determinism in which many alleles act in combination to produce a phenotype.

Polymorphism A relatively common genetic variant. Usually defined as a genetic variant for which the rarer allele frequency is greater than 1 percent.

Posterior probability The probability assigned to the validity of a proposition after information has been gathered (see *Bayesian reasoning*, *Prior probability*).

Primer A short fragment of DNA used to initiate the replication of a specific DNA target sequence.

Prior probability The probability assigned to the validity of a proposition before further information is gathered (see *Bayesian reasoning, Posterior probability*)

Profiling The use of a combination of markers for prediction.

Psychopathology A psychiatric disease or manifestation of psychiatric disease.

Random error Error occurring at unpredictable intervals, and with unpredictable effects.

Receptor A protein that recognizes a small molecule, for example a neurotransmitter or an endocrine hormone, and initiates a signal.

Recessive Requiring genotypes with two copies of an allele to manifest a trait.

Red Queen hypothesis The proposal that species have to evolve as rapidly as possible, just to stay in the same place on the evolutionary treadmill.

Resiliency The ability to withstand mental and physical stress.

Ribosomes Cellular organelles whose function is the translation of RNA into protein.

RNA Ribonucleic acid. A single-stranded molecule which can form complex structures due to internal base pairing. Usually comprised of adenine (A), guanine (G), uracil (U) and cytosine (C) nucleotides, carrying the genetic code. RNAs are translated into proteins and also have independent structural and regulatory functions.

Schizophrenia A relatively common psychiatric disease marked by thought disorder, diagnosed operationally by criteria of duration and problems in symptom areas.

Serotonin A usually inhibitory neurotransmitter ultimately derived from tryptophan, an amino acid found in the diet.

Serotonin transporter Terminates the action of serotonin in the synapse by uptake.

Sex Genetic sexual identity (see *Gender*).

Skinner box A chamber in which a subject may be presented with defined stimuli and conditioned to associate these with particular outcomes.

Smallpox A severe, frequently fatal or disfiguring viral illness, now eradicated except for laboratory cultures.

SNP Single nucleotide polymorphism. A polymorphism involving a single DNA nucleotide substitution, for example A to G, T or C. SNPs usually have two alleles.

Social Darwinism The belief that higher social status was the product of superior genotype.

Socialism A political system involving sharing.

Stop codon In DNA or RNA, a three-nucleotide word that instructs the ribosome to terminate translation of the RNA.

STR Short tandem repeat. A polymorphism involving different numbers of repeats of a certain DNA sequence, which can range from one nucleotide to many nucleotides in length. STRs often have many alleles.

Stroop test A neurocognitive test of frontal lobe function that accesses the subject's ability to switch strategies by which it comprehends stimuli.

Systematic error An error intrinsic to a process and creating a predictable bias.

Trait Any measurable characteristic of an organism that is relatively stable. Some traits are heritable (see *Phenotype, Intermediate phenotype, Endophenotype*).

Transcription The enzymatic copying of DNA into an RNA of complementary sequence.

Translation The synthesis of a protein directed by an RNA sequence.

Warrior/worrier model A selectionist explanation for the maintenance of COMT alleles that have counterbalancing effects in cognition versus resilience to stress and pain.

Wisconsin Card Sort Test (WCST) A neurocognitive test of frontal lobe function that requires the subject to switch strategies needed to match cards to a target.

X-linked Due to a gene carried on the X-chromosome so that distinct patterns of transmission may be seen depending on whether the gene is recessive (in which case male offspring of normal female carriers are seen) or not.

Index

Edwards Brothers Malloy
Ann Arbor MI. USA
May 24, 2013